pH Measurements

pH Measurements

C. Clark Westcott

BECKMAN INSTRUMENTS, INC.
IRVINE, CALIFORNIA

1978

ACADEMIC PRESS   New York   San Francisco   London

A Subsidiary of Harcourt Brace Jovanovich, Publishers

ACADEMIC PRESS, INC.
111 Fifth Avenue, New York, New York 10003

United Kingdom Edition published by
ACADEMIC PRESS, INC. (LONDON) LTD.
24/28 Oval Road, London NW1 7DX

Library of Congress Cataloging in Publication Data

Westcott, C   Clark.
    pH measurements.

    1. Hydrogen-ion concentration—Measurement.
I. Title.
QD561.W52   541'.3728   77-11227
ISBN 0-12-745150-1

PRINTED IN THE UNITED STATES OF AMERICA

Contents

Preface

This book is designed to be used for solving problems or obtaining a high degree of accuracy in practical pH measurement. It provides a simplified presentation of the subject.

The operator of pH equipment may be interested in the results, but knowing how to make a pH measurement is a prerequisite. Users of pH equipment cannot afford to ignore this critical parameter because a small deviation in pH can mean the difference between the success or failure of a process.

A pH measurement appears to be a simple operation. Indeed, there are only three main components involved. However, data taken as pH measurements may appear to be nonsense. The ability to interpret the results becomes a matter of understanding the basics of pH measurement.

I have dealt with pH measurement for more than a decade and have found that difficulties are usually experienced with specific practical applications. As with many analytical measurements, there is normally a common thread to the problem and its solution, and to find this thread the operator must understand the basics of pH measurement.

The first chapter deals with the basic theory of pH. It is not an involved academic discussion, but rather an explanation of how the parameters that influence pH measurements can affect the results.

The next three chapters describe the characteristics, care, and performance of pH equipment and standard solutions. The discussions are directed toward providing long-term practical pH measurements through a better understanding of each component.

The last three chapters contain information concerning the use of proper techniques for difficult applications. They will be particularly useful to a person being trained or wanting to learn how to make pH measurements. The application examples are those with which I am most frequently confronted, but they are discussed as a general approach to taking pH measurements in difficult samples. The last of these chapters suggests possible sources of difficulties, how to locate them, and possible solutions.

I would like to thank Dr. George Matsuyama for his patience and help in the writing and editing of this book.

pH Measurements

Chapter 1

Principles of pH Measurements

The objective of this chapter is to relate the principles of pH measurements to the actual measurement. In other words, what are the factors involved in the measurement and what magnitude of effect does each factor have on the measurement? The chapter is not intended as a theoretical discussion, and for deeper under-standing of the theory and thermodyamics involved, other refer-ences are suggested [1–3].

1.1   DEFINING pH

The principles of pH begin with a definition of the term pH. The p comes from the word power. The H, of course, is the symbol for the element of hydrogen. Together, the term pH means the hydrogen ion exponent.

The pH of a substance is a measure of its acidity just as a degree is a measure of temperature. A specific pH value tells the exact acidity. Rather than stating general ideas such as orange juice is acid or the water is hot, a specific pH value gives the same relative point of reference, thus providing more exact communication. The orange juice has a pH of 4.0 or the water is at 80°C provides an exact common language.

pH is defined in terms of the hydrogen ion activity:

$$pH = -\log_{10} a_{H^+} \qquad \text{or} \qquad 10^{-pH} = a_{H^+} \qquad (1)$$

pH equals the negative logarithm of the hydrogen ion activity, or the activity of the hydrogen ion is 10 raised to the exponent $-pH$. By the latter expression, the use of the p exponent becomes more obvious. The activity is the effective concentration of the hydrogen ion that is in solution. It is discussed in more detail in Section 1.2. Basically the difference between effective and actual concentration decreases when moving toward more dilute solutions in which ionic interaction becomes progressively less important.

The formula for pH is analogous to the relationship between absorbance (A) and transmittance (T); that is, $A = -\log T$. In this log function, however, the range is normally much narrower than that for pH. The important similarity is the logarithmic relationship; that is, for every decade change in activity, the pH changes by one unit. The scope of this relationship is illustrated in Figure 1.1. The factor of 10 between each pH unit shows the importance of being able to measure pH to a tenth or a hundredth of a unit.

Normally, reference is made to the hydrogen ion when reference should be made to the hydronium ion (H_3O^+). It is a matter of convenience and brevity that only the hydrogen ion is mentioned even though it is normally in its solvated form:

$$H^+ + H_2O \rightleftharpoons H_3O^+ \qquad (2)$$

The complexing of the hydrogen ion by water is a factor which affects activity and applies to other ions which partially complex or establish an equilibrium with the hydrogen ion. In other words, equilibriums such as

$$H_2CO_3 \rightleftharpoons H^+ + HCO_3^- \qquad (3)$$

$$HC_2H_3O_2 \rightleftharpoons H^+ + C_2H_3O_2^- \qquad (4)$$

complex the hydrogen ion so that it is not sensed by the pH measuring system. This, of course, is why an acid–base titration is

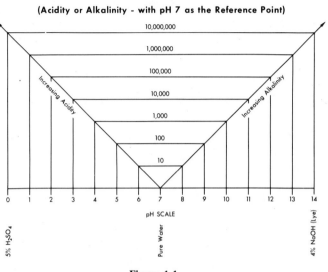

Figure 1.1
pH Scale Logarithmic Relationship

performed if total concentration of acid (H$^+$) is desired. These effects on hydrogen ion activity are obvious, but other more subtle effects are involved in the correlation of activity and concentration.

1.2   ACTIVITY VERSUS CONCENTRATION

Since the pH glass electrode is sensitive to hydrogen ion activity a_{H^+}, the factors which influence activity and its definition are of primary importance. The activity of the hydrogen ion can be defined by its relation to concentration (C_{H^+}, molality) and the activity coefficient f_{H^+} :

$$a_{H^+} = f_{H^+}C_{H^+} \tag{5}$$

If the activity coefficient is unity, then activity is equal to concentration. This is nearly the case in dilute solutions, where the ionic strength is low. Since the objective of most pH measurements is to find a stable and reproducible reading which can be correlated to the results of some process, it becomes important to know what influences the activity coefficient and therefore the pH measurement.

The factors that affect the activity coefficient are the temperature T, the ionic strength u, the dielectric constant ϵ, the ion charge Z_i, the size of the ion in angstroms $\mathring{a}$, and the density of the solvent d. All of these factors are characteristics of the solution which relate the activity to the concentration by two main effects. The first is the salt effect designated as $f^x{}_{H^+}$. It can be approximated for the hydrogen ion by the expression

$$\log f^x{}_{H^+} = \frac{-0.5u^{1/2}}{1 + 3u^{1/2}} \tag{6}$$

where u is the ionic strength which is defined as one half the sum of molality times the square of the charge of the ionic species:

$$u = \tfrac{1}{2} \sum C_m Z_i^2 \tag{7}$$

The more exact definition of this salt effect is found by using the Debye–Hückel equation described in the Glossary. The other factors mentioned are used in defining this equation, thus showing their effect on the measurement.

Examples of the salt effect can be shown by using the approximation equation (6) at different molalities for the hydrogen ion. If a monovalent anion and hydrogen ion are assumed, the charge factor Z_i reduces to one, and the molality is the main factor in calculating the ionic strength. The approximate influence of the salt effect on the activity coefficient for different molalities can be seen in Table 1.1.

TABLE 1.1

Approximate Salt Effects, $f^x{}_{H^+}$

Molality:	0.001	0.005	0.01	0.05	0.1
Activity coefficients:	0.964	0.935	0.915	0.857	0.829

An example of a salt effect can be illustrated by the following:

$10^{-2}\ m$ HCl solution

$$pH = -\log(C_{H^+} f^x{}_{H^+})$$
$$= -\log(0.01 \times 0.915)$$
$$= -\log(9.15 \times 10^{-3})$$
$$= 2.04$$

$10^{-2}\ m$ HCl solution plus $0.09\ m$ KCl added

$$pH = -\log(0.01 \times 0.829)$$
$$= -\log(8.29 \times 10^{-3})$$
$$= 2.08$$

In other words, the pH is increased by 0.04 pH unit or the activity is decreased in the higher ionic strength solution. Therefore, samples with the same hydrogen ion content can expect to have different pH values if the ionic strength of the sample varies.

The second effect is the medium effect which is designated as $f^m{}_{H^+}$. This effect relates the influence that the solvent will have on the hydrogen ion activity. It reflects the electrostatic and chemical interactions between the ion and the solvent, of which the primary interaction is solvation. This effect can be related by comparing the standard free energy in a nonaqueous solvent to that in water. For example, the activity of hydrogen ion in ethanol is much greater (≈ 200 times) than in water.

This brings up the question about nonaqueous pH measurements which are covered in more detail in Chapter 6. Most often an aqueous pH buffer solution is used to standardize the pH measuring system. If the measurement is to be made on a nonaqueous sample, the correlation between the activity of hydrogen ion in an aqueous standard and the activity in a nonaqueous sample is not valid. If, however, the pH value obtained is stable and can be correlated to some results, the hydrogen ion activity need not be known. The relative pH value can be used as an indicator to alter the process or to proceed in some corrective manner if the pH value changes dramatically. In other words, when defining pH, normally aqueous samples are implied and the exact activity of hydrogen ion in nonaqueous sample is unknown (see Section 4.4.3).

Thus, the activity is related to concentration through a salt effect and a solvent effect. The glass electrode measurement of activity is mainly influenced by the ionic strength, the temperature, and the solvent:

$$a_H = f^x{}_{H^+} f^m{}_{H^+} C_m \tag{8}$$

This means that the sample composition and conditions should be stated when stating the pH value if another person is to duplicate the results or if pH values are going to be compared. The pH of the solution is valid only at a particular temperature, ionic strength, and stated solvent.

Because of these influences, a sample pH value cannot be extrapolated to another temperature or dilution. If the pH value of a particular solution is known at 40°C, it is not automatically known at 25°C. The standard buffer solutions were studied at different temperatures and compositions to define their activity, and unless the same is done for a sample, its pH under different conditions is not known because of these variables.

1.3 pH SCALE

The pH scale was established to provide a convenient and effective means of communication with regard to the relative acidity and basicity of a particular solution. Its range is based on the dissociation constant for water, K_w ($K_w = a_{H^+} \cdot a_{OH}$). In pure water, hydrogen ion (H^+) and hydroxyl ion (OH^-) concentrations are equal at 10^{-7} M at 25°C. This is a neutral solution. Since most samples encountered will have less than 1 M H^+ or OH^-, the extremes of pH 0 and pH 14 are established. Of course, with strong acids or bases, pH values below 0 and above 14 are possible but

TABLE 1.2

Temperature versus pH Scale

Temperature (°C)	$-\log K_w$	K_w
0	14.943	1.14×10^{-15}
10	14.535	2.9×10^{-15}
20	14.167	6.8×10^{-15}
25	13.996	1×10^{-14}
30	13.833	1.47×10^{-14}
40	13.535	2.9×10^{-14}
50	13.262	5.47×10^{-14}
60	13.017	9.6×10^{-14}

infrequently measured. The change in the dissociation constant K_w with temperature affects the pH at which neutrality is obtained and the pH of the basic solution (see Table 1.2). This change has little effect on the pH of acidic solutions.

Table 1.3 illustrates that a neutral solution or a 1 M OH⁻ solution has a different pH value depending on the temperature, and this effect on the pH value increases with greater alkalinity. Also, the deviation from pH 7.0 as being neutral or pH 14 being 1 M OH⁻ increases with the deviation from 25°C at which the 0–14 pH scale is symmetrical around neutrality.

TABLE 1.3

Neutral or Basic Solutions versus Temperature

Solution	Temperature (°C)		
	25°	0°	60°
Neutral			
(H^+)	10^{-7}	3.3×10^{-8}	3.1×10^{-7}
(OH^-)	10^{-7}	3.3×10^{-8}	3.1×10^{-7}
pH	7	7.47	6.51
Basic			
(H^+)	10^{-14}	1.14×10^{-15}	9.6×10^{-14}
(OH^-)	10^{0}	10^{0}	10^{0}
pH	14	14.943	13.017

The concentration of the hydrogen ion as interpreted from pH values varies with temperature in neutral to basic solutions. When the hydrogen ion concentration is not the dominant ion, the equation for pH can be written as

$$pH = -\log \frac{K_w}{(OH^-)} \qquad (9)$$

Example

$40°C$ measured pH 11.88, what is the (OH^-)?

$$11.88 = -\log 2.9 \times 10^{-14} + \log OH^-$$
$$\log OH^- = 11.88 - 13.535 = -1.655$$
$$OH^- = 2.2 \times 10^{-2} \quad \text{versus} \quad 7.6 \times 10^{-3} \qquad \text{if} \quad 10^{-14}K_w \text{ is used.}$$

Converting pH readings to concentration (activity) and vice versa is shown in the following examples.

$$pH = -\log(\text{hydrogen ion activity})$$
$$pH = -\log(a_{H^+})$$

Examples

pH to concentration

$$pH = -\log(3 \times 10^{-4})$$
$$= -(+0.477 -4)$$
$$= 3.523$$

Concentration to pH

$$pH = -\log(a_{H^+})$$
$$\log(a_{H^+}) = -pH$$
$$= -3.523$$

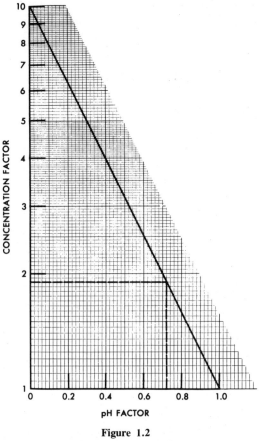

Figure 1.2
pH versus Concentration

pH scale conversion: The pH factor provides the decimal value the same as the mantissa would using log tables. The integer value is supplied by the exponent. For example, a concentration of 1.9×10^{-7} gives a factor of 0.72 by extrapolating from the concentration axis to the pH factor axis. The integer value is one less than the exponent, 7, and therefore the pH is 6.72.

Mantissa converted to positive value

$$(a_{H^+}) = \text{antilog}(+0.477 - 4)$$
$$= 3 \times 10^{-4}$$

A calculator with log functions or a simple four-place log table are convenient means of conversion. Also, a piece of semilog paper may be used with only slightly less precision. An example using semilog paper is shown in Figure 1.2.

1.4 pH MEASURING SYSTEM

The activity of the hydrogen ion in solution is measured with a pH measuring system consisting of a glass electrode, a reference electrode, and a pH meter. Each component is described in detail in the following chapters. The purpose of this section is to show how each component relates to the actual measurement.

When the pH-sensitive glass bulb is immersed in a solution, an exchange equilibrium is established between the hydrogen ion and the ions in the glass. This equilibrium is the source of the potential measured. The potential which is measured varies with the hydrogen ion activity in a known manner. The glass electrode alone, however, is not sufficient to measure the potential, since a reference electrode is needed to complete the measuring circuit.

The reference electrode supplies a stable reference potential against which the potential from the glass electrode may be compared. The reference electrode provides a stable potential by surrounding an internal element with a known solution. For example, a calomel internal is surrounded by saturated potassium chloride filling solution to provide 244 mV versus the hydrogen electrode. The filling solution makes contact with the sample solution through a junction to complete the circuit.

Thus, the glass electrode potential E_g is proportional to the hydrogen ion activity a_{H^+}, and the reference potential E_r is proportional to a standard potential E^0 which is dependent on the type of internal. This can be represented by the expressions

$$\text{Glass:} \qquad E_g \propto a_{H^+} \text{ (pH)}$$

$$\text{Reference:} \quad E_r \propto E^0$$

$$\text{Difference:} \quad E_r - E_g \propto E^0 - a_{H^+}$$

In the measuring system, the glass potential is compared to the reference potential, and the difference between these potentials is the observed potential. This is a simplified form of the observed potential because there are actually many potentials involved as shown in Figure 1.3.

The glass electrode internal wire establishes a potential with the solution inside the glass bulb (E_1). There is a potential established between the internal solution and the inside glass surface (E_2). The potential represented as E_5 in Figure 1.3 is the potential established by the hydrogen ion activity. The reference electrode internal potential is represented as E_9, and the reference junction potential is represented as E_7. It is this latter potential E_7 that will be discussed in detail since it is the cause of most problems encountered when making a pH measurement. Hopefully, all of the potentials, except E_5, are stable and reproducible so that the only variable is the potential established by the hydrogen ion activity.

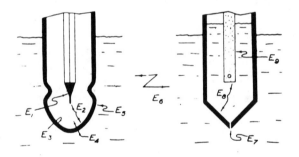

Figure 1.3
Sources of Potentials

Potential	Between
E_1	Internal and solution
E_2	Internal and inside glass membrane
E_3	Solution and inside glass membrane
E_4	Insider and outer glass membrane
E_5	Sample and reference electrodes
E_6	Glass and reference electrodes
E_7	Liquid junction and sample
E_8	Internal and filling solution
E_9	Internal elements

As previously mentioned, the glass electrode potentials are compared to the reference electrode potential and the difference displayed on the pH meter. When immersed in pH 7 buffer solution, most pH glass electrodes are designed to provide a potential equal to that of a saturated calomel reference electrode (SCE). Thus, at pH 7, the difference in potential is approximately zero. As the solution becomes more acid, the glass potential becomes greater (more positive millivolts) than the reference electrode potential, and as the solution becomes more alkaline, the glass potential becomes less (more negative millivolts) than the reference electrode potential. This is illustrated in Figure 1.4.

The third component of the pH measuring system is the pH meter. The glass electrode bulb has a high resistance across it of about 100 megohms. This fact prevents the use of an ordinary voltmeter for reading out the electrode potentials. A meter which has a high input impedance or low bias current is required. The pH meter also provides many special functions besides the readout, such as the temperature compensation, all of which are discussed in Chapter 2. Figure 1.5 illustrates the complete system.

As stated previously, the observed potential from the glass electrode varies in a known manner, the Nernst equation. This equation shows the relationship between any potentiometric sensing electrode and the ion to which the electrode is sensitive. When applied to the pH glass electrode and the hydrogen ion activity, it

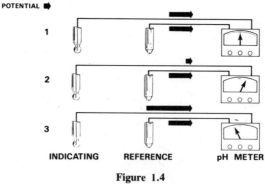

Figure 1.4
Comparing Potentials

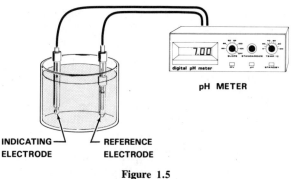

Figure 1.5
pH Measuring System

may be represented in a simplified form as

$$E_{obs} = E^{0'} + S \log a_{H^+} \tag{10}$$

where

E_{obs} is the observed potential,
$E^{0'}$ the stable fixed potentials including reference internal
   potential, and
S the slope factor.

Since pH has already been defined as $pH = -\log a_{H^+}$, substitution
into equation (10) provides the relationship between the observed
potential and the hydrogen ion activity. This is

$$E_{obs} = E^{0'} - S \quad pH \tag{11}$$

The more complete form of the Nernst equation is given in the
Glossary, but one major component of the slope S needs expansion
at this time; that is, the slope varies with temperature. Equation
(11) may be written as

$$E = E^{0'} - 0.198T_k \quad pH \tag{12}$$

where T is the temperature in degrees Kelvin. Therefore, the
amount of potential observed from the electrode will vary with

temperature. This is another way in which temperature affects the pH besides its effect on the activity and the conversion from pH to concentration. This equation also shows that if temperature is changing, the potential is changing, thus requiring that the electrodes be at thermal equilibrium with the solution in which the tips are immersed before a stable pH reading can be obtained. The magnitude of the slope is detailed in Table A.1, and representative values are listed in Table 1.4.

As mentioned previously, a pH glass electrode and a calomel reference electrode immersed in pH 7 buffer generate approximately zero millivolts. The standardization or calibration control on the pH meter allows any deviation of these electrodes from zero millivolts to be offset, thus providing comparison or standardization of all electrode pairs. This provides a starting point for a linear function which represents the observed potential versus the pH, and since the slope of this function is predictable, it can be represented by the function illustrated in Figure 1.6.

At 25°C each additional pH unit represents 59.16 mV, and at 100°C each additional pH unit represents 74.04 mV from the starting point of pH 7. The pH meter and electrodes are designed with a value of pH 7 being approximately zero millivolts. This provides smaller potential measurements over the 0 to 14 pH range than if zero millivolts were observed at pH 0 for example, and also because the temperature response of the glass electrode (discussed in Section 2.3.1) is a function of this point. The temperature coefficient (dE/dt) of the glass electrode is normally designed to

TABLE 1.4

Temperature versus Potential

Temperature (°C)	Slope (mV/pH unit)
0	54.20
25	59.16
37	61.54
60	66.10
100	74.04

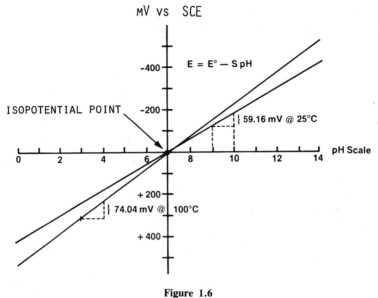

Figure 1.6
Potential versus pH

match the calomel reference electrode temperature coefficient to
the isopotential point, for either electrode is approximately the
same. As the pH increases beyond pH 7, more negative millivolts
are observed. As the pH decreases below pH 7, more positive
millivolts are observed. Thus at 30°C, where S equals about 60 mV/
pH units, pH 0 is represented by +420 mV and pH 14 is
represented by −420 mV.

Figure 1.6 also shows that near pH 7 only a slight change in pH
is caused by a change in the slope (temperature compensator on
pH meter), and as the extremes of the pH scale are approached,
the effect or change in the pH value becomes greater. In fact, with
a pair of electrodes that produce zero millivolts in pH 7 buffer
solution, there should be no change in pH when the temperature
compensator is rotated. Thus, if a measurement is being made in a
solution which is close to neutral, the accuracy of the temperature
compensator is much less important than in a solution not near
neutral pH.

TABLE 1.5

pH versus Percent Error

Error (mV)	pH	$\Delta C/c$ (%)
1	0.017	3.9
4	0.068	15.6
16	0.270	62.3
32	0.541	124.6

1.5 POTENTIAL ERROR

Another consideration of the logarithmic function of pH is the pH or percent concentration error resulting from a potential measurement error. The previously stated equation (11) represents the potential measured by a pair of electrodes. The difference between two measurements can be represented as

$$\frac{E_1 - E_2}{S} = pH_2 - pH_1 = \log \frac{C_1}{C_2} \tag{13}$$

or

$$\frac{\Delta E}{S} = \Delta pH = \Delta \log C$$

If the two samples are identical or in other words $C_1 = C_2$, but a measurement error resuts, the error in pH value or percent concentration change can be seen in the data presented in Table 1.5. It can be seen from these data that a small millivolt error represents a large percentage difference in concentration.

References

1. Bates, R. G., "Determination of pH, Theory and Practice." Wiley, New York, 1973.
2. Lingane, T. T., "Electroanalytical Chemistry," 2nd ed. Wiley (Interscience), New York, 1958.
3. Furman, N. H., "Treatise on Analytical Chemistry," Part I, Section D-2, Potentiometry, p. 2269. Wiley, New York, 1963.

Chapter 2

The pH Meter

This chapter describes how a pH meter functions and how its various controls affect the pH reading. The types of readout and the effects of temperature compensation and standardization controls are discussed. Answers are given to questions such as when to use a slope control or how a zero control relates to the slope adjustment. In conclusion, the various types of meters and their performance specifications are discussed.

2.1 BASIC pH METER CIRCUITRY

Before the external controls are described, a basic understanding of the pH meter circuitry is needed. The pH-reference electrode pair acts as a battery when immersed in solution. The voltage developed at the electrodes is measured by applying it to a high-impedance input amplifier. The amplifier output is modified by the various controls on the pH meter. The slope and temperature compensation controls affect the amplifier gain, thus compensating for changes in electrode voltage output caused by variations in interference and temperature. The pH meter may have a zero or isopotential control which allows setting zero-millivolt readout at values within some pH range other than the normal pH 7 value. This allows, for example, pH 4 to correspond to a zero-millivolt readout even though the electrodes are producing approximately 180 mV versus SCE. This control is used in conjunction with the slope control as discussed in Section 2.4. The output from the amplifier is fed to the readout network in order to display the appropriate pH value. The amplifier output may also be fed to a recorder to allow pH values to be recorded.

The reference electrode is connected to the standardization control. Thus the potential of the reference electrode and the potential of the standardization control together offset the glass electrode potential to provide an adjustable displayed potential. This is possible since changing the standardization potential changes the overall potential observed by the amplifier (see Figure 2.1).

One specification often stated is the pH meter input impedance. Its value is typically 10^{13} ohms. This high impedance value is required to handle the high resistance of the glass bulb. The glass bulb resistance value typically varies with electrode configuration and type of glass, but is usually between 10^6 and 10^9 ohms. The

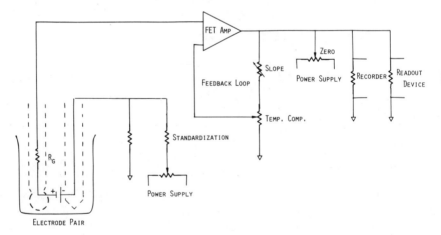

Figure 2.1
Basic pH Circuit

TABLE 2.1

Impedance Error

R_m	×	$I_b =$	E (mV)	pH error (unit pH $\cong$ 60 mV)
10^8		1 picoamp	0.1	0.002
10^9		1 picoamp	1.0	0.02
10^8		10 picoamp	1	0.02
10^9		10 picoamp	10	0.2

closer the input impedance value is to the glass resistance, the more error is possible. Therefore, for applications in which a high resistance sample (e.g., nonaqueous solutions) or a high resistance glass bulb are used, it is important to use a high input impedance pH meter. The electronic specification most important in determining input impedance is the bias current. This specification represents the actual system performance rather than being a resistance value. In other words, how much current leakage is generated in the amplifier circuit? This current value will determine the potential error observed when the electrode impedance changes. Its typical value should be around a picoampere (10^{-12} amp). If, for example, the measuring system resistance is represented by R_m and the bias current is represented as I_b, the resulting pH error from varied values of these parameters is noted in Table 2.1. The bias current and electronics will vary with temperature and can induce a significant error in the system. Therefore, temperature-stable circuit and components must be employed.

As previously mentioned, a pH meter is designed to provide zero millivolts at pH 7, and increasing positive potentials represent lower pH values while increasing negative potentials represent higher pH values. The glass and reference electrodes are also designed to have the same potential and temperature coefficient in pH 7 buffer solution. This is called the electrode isopotential point. Theoretically it is a point at which the same potential is observed regardless of the temperature as shown in Figure 1.6. In practice, however, the electrodes do not provide a point, but an area is observed as shown in Figure 2.2.

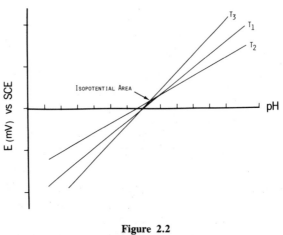

Figure 2.2
Isopotential Area

When electrodes are manufactured, every practical attempt is made to minimize this area. Electrodes that do not meet an asymmetry potential specification are eliminated. The asymmetry potential, in this case, is any difference in potential between glass and SCE reference electrodes, when immersed in pH 7 buffer (see Section 3.1.4 for an exact definition). Since it is not possible to obtain an isopotential point with electrodes, the pH meter isopotential point is set at the most likely point, pH 7. Since this point is only an estimation of the electrodes isopotential point, a slight error is observed if the measuring temperature is different than the buffering temperature. This is because a change in the slope made by changing the temperature compensator of the meter may not revolve around the same point as the slope of the electrode pair.

2.2   READOUT

The two most common pH readout displays are analog and digital. Each has advantages, and the selection of the type of readout should be based on which of the advantages best suits the application.

2.2.1 Digital

The digital display has elements of simplicity and exactness that are not found on a scale and needle readout. The exact pH value is clearly displayed without need of interpretation. The pH value is always in expanded form reading to a hundreth or a thousandth of a pH unit without changing modes. The exact pH value is easily recorded on paper tape by connecting the BCD (binary coded decimal) output from a digital pH meter to a printer.

Once the digital meter is calibrated, its pH range is not limited by a scale for expanded readout. For example, on some analog meters calibrated for an expanded range of pH 6 to pH 8, a recalibration may be required if a sample with pH 8.4 is encountered. Another limitation of a meter with a scale is the measurement of pH values greater than the scale of 0 to 14. If strong acids or some nonaqueous solvents are to be measured at pH values below 0 or above 14, a standardization offset potential and difficult interpretation of the pH scale would be required on an analog meter. Most digital meters display from $+19$ to -19 pH units, although only part of the range is useful.

These advantages would be particularly useful where numerous pH measurements are being taken with a requirement for reliable values over a wide pH range.

2.2.2 Analog

The analog meter has the advantages of concentration readout, indication of rate, and high reliability. Direct concentration readout is possible by providing a log scale. This is particularly useful when using ion-selective electrodes where a pCation or pAnion scale is not yet established. For example, an analog meter can be calibrated to 10 ppm Na^+ and then a slope adjustment made to make the potential from a 100-ppm Na^+ solution indicate at the appropriate point on the log scale. Any sample within this range can then be read directly on the log scale. This eliminates the construction of a semilog calibration curve since this scale is provided on the meter.

The rate of pH change can be more easily interpreted on an analog meter than on the digital display. The needle travel per period of time is a better indication of pH rate change than changing digits. This is particularly useful when performing titration, and increments of titrant cause a pH change.

The last advantage is fast fading, but when the digital displays first became available they did not have the reliability of an analog meter. This is understandable since the analog meter has been in use for a longer period of time and has gained reliability with use. A summary of the advantages for each readout are shown in Table 2.2.

Whether an analog meter or a digital display is used, the meter reliability should not be the limiting factor on the accuracy requirement. Most often the electrodes or the operational technique are the limiting factors on accuracy. The greater the accuracy requirement, however, the greater the number of parameters to be considered that affect the accuracy. For example, a meter with ±0.001 readability may seem unnecessary when a buffer accuracy is ±0.005 pH or a degree change in temperature causes about 0.003 pH unit change if a pH unit from standardization. But the thousandth of a pH readability can be used as a reliability check on the hundreth of a pH unit. For example, if the readout indicated pH 7.017, it could be displayed as 7.01 on a display which reads only to the second place. In general the display readability should be from five to ten times greater than the required accuracy. An analog meter that reads to ±0.002 can provide an answer as accurate as a digital display that indicates ±0.001 pH unit.

The pH meter recorder output is another method of obtaining expanded readability, permanent record, response time, or pH rate changes. Although measurement accuracy is not increased, the readability of the pH displayed can be expanded on a recorder.

TABLE 2.2

Advantages of Digital versus Analog Readout

Digital	Analog
Clarity of pH value	Indicating rate of change
Wider pH scale (<0 and >14)	Log scales
BCD output	
No recalibration	
Always expanded readout	

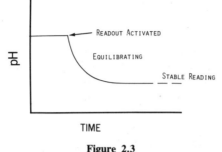

Figure 2.3
Recording pH versus Time

Under these conditions more noise is usually observed since the amplifier is selected for the stability required for the standard readout. The recorder display output is also an excellent method of determining rate of pH change with time or when a stable reading is obtained (see Figure 2.3). It should be utilized for greater precision and convenience.

2.2.3 Automatic Readout

A new aspect to the readout has recently become available on some commercially available digital pH meters. This is a feature which provides the operator with an indication of when a stable reading is obtained. It is similar to the response curve on a recorder in that it will indicate when a stable pH reading is observed over a period of time. It functions by comparing the initial pH value at some initial time with the pH value after a set time interval, and then calculates the pH deviation over that time period. It removes the guesswork from determining when to record a pH value.

2.3 TEMPERATURE

As shown in the basic circuit diagram, Figure 2.1, the temperature compensation control changes the output slope (mV/pH unit) to correspond to the Nernst equation factor. The slope value can be calculated from equation (12), where $0.1984T_k$ represents the slope. Examples of a slope value versus temperature are given in Table 2.3 with a full listing in Table A.1.

The right-hand column expresses the magnitude of error observed because of the change in slope if the compensator is set improperly or out of calibration. In other words, the meter compensation is different from the electrode potential with temperature change. An average value of 0.003 pH error/°C error at 1 pH unit from standardization can be used to predict a magnitude of error. Figure 1.6 illustrates the change in slope (mV/pH unit) with respect to pH.

When a meter is set at pH 7 and has zero millivolts on its input, the temperature compensator has no effect on the reading. It is only as the reading potential becomes greater or less than zero millivolts that the temperature compensator has an effect. Therefore, as the extremes of the pH scale are approached, the greater the effect the temperature compensator has on the reading. An operator should be more concerned with the temperature compensator setting in these areas.

If the temperature compensation control is left at the standardization temperature and the sample is at a different temperature, the correct sample pH can be calculated with the formula

$$pH = 7 - \frac{T_1}{T_2}(7 - pH_0)$$

where

pH$_0$ is the pH observed,
T_1 the temperature of standardization (°K), and
T_2 the temperature of the sample (°K).

For example, if a pH measuring system was standardized at 298°K (25°C) and a pH of 10.0 was observed without adjusting the temperature for a sample which was at 273°K (0°C), the correct pH can be calculated as

$$pH = 7 - \frac{298}{273}(7 - 10) = 10.275$$

TABLE 2.3

Slope versus Temperature

Temperature (°C)	Slope (mV/pH)	pH error/°C error with reading 1 pH from standardization
0	54.196	0.0037
30	60.148	0.0033
60	66.100	0.0030
100	74.036	0.0027

It is wise to keep in mind the magnitude of error which can be caused by temperature change since it is often less than the required accuracy or the error being observed.

2.3.1 Isopotential Point

The temperature compensator is keyed to the zero-millivolt isopotential point of the meter. Since this point may be varied with a zero control (see Section 2.4), the compensation will vary with this control setting. The electrodes, on the other hand, are fixed as to their isopotential point and cannot be varied. It is only when the isopotential point of electrodes and the meter are identical, or nearly so, that temperature compensation can be applied. In other words, if slope correction is being applied with the zero control at other than pH 7, the standard buffer solution and sample should be at the same temperature since the temperature compensator does not apply the proper correction. (See Figure 2.4.)

Both automatic and manual temperature compensation are provided on most pH meters. The automatic temperature compensation (ATC) is obtained through the use of a thermistor or wire-wound resistance probe. The probe resistance value changes with temperature, and the circuitry in the pH meter uses the probe resistance value to adjust the amplifier gain to the appropriate slope value (mV/pH). An ATC probe is useful when long-term pH monitoring is being performed on a sample which may have slight temperature variations. Therefore, it is usually employed when an industrial process stream is being monitored. If, however, quick

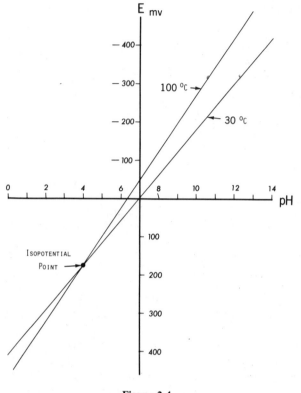

Figure 2.4
Isopotential Point

accurate laboratory measurements are being made, the ATC probe is not as useful. The probe requires time to come to equilibrium with the sample, which may be greater than the time required for electrodes to equilibrate. It also may not be as accurate as a manual temperature compensator set at a thermometer temperature. In other words, the resolution of the thermistor to a temperature change may be less than that of the manual temperature compensator. The ATC probe may be used in a dual role to add greater utility to the pH meter. It may be used as a thermometer.

Some manufacturers supply the temperature readout option with some models of digital pH meters.

2.4 CALIBRATION

The pH meter is calibrated using either a single-point standardization with 100% slope or a two-point calibration with the first point for standardization combined with the second point for a span adjustment through use of a slope control. If all the sample pH values are close to the point of standardization, such as would be the case for blood pH measurements, there is little value in making a span adjustment. If, on the other hand, the sample pH values are over a pH range and a high degree of accuracy is required, a slope adjustment should be made. This is particularly true at high pH values where nonlinearity of electrode response is more likely to be encountered. The pH meter slope control performs the same type of function as the temperature compensation control; that is, it changes the slope (mV/pH) of the meter output revolving about the zero-millivolt point as shown in Figure 2.4.

The purpose of the slope and zero (isopotential) controls is to provide a pH readout that closely follows the electrode response, thus increasing the accuracy of the measurement. If a high degree of accuracy is not required, typically greater than ±0.05 pH unit or more, the slope control should be turned off or set at 100% and the zero control set to display the typical isopotential point at pH 7 and no second slope/span adjustment made.

If the increased accuracy of a slope adjustment is required, the two-point calibration should be made over a narrow pH range so that the meter response will closely approximate the electrode response.

In the simplest case with a single-point standardization, the pH meter standardization control is used to adjust any deviation of potential of the electrode pair from the ideal Nernst response. If all electrodes produced the same potential for a known buffer solution and did not change with time, there would be no need for a standardization control. Since the response of most electrodes is not ideal, the closer the sample pH value is to the standardization point, the less error will be observed.

If a wide-range slope adjustment is made, for example, with the initial zero-millivolt point at pH 7 and the slope adjustment made at pH 13, the electrode pair is supposedly calibrated over this range. Suppose the electrode pair is standardized in pH 7 buffer solution and when immersed in pH 13 buffer solution, the display indicates 12.2; the slope adjustment would be made to make the display read pH 13.0. The number of millivolts per pH unit is lowered by the slope adjustment, since the electrode pair is providing fewer millivolts than the ideal Nernst output. By making the slope adjustment, the error due to the short span of the electrode pair is assumed to be linear (line $\overline{AD}$ is moved to $\overline{AC}$ in Figure 2.5); that is, the deviation of 0.8 pH unit is assumed to be spread linearly over the pH 7 to pH 13 range. The operator knows that the response at pH 7 and at pH 13 is correct, but does not know about pH values within this range.

A nonlinear type of electrode response (curve $\overline{ABC}$, Figure 2.5) is more likely. This curve is accentuated for illustrative purposes. When the slope adjustment is made, the nonlinearity remains with only the two points being on the correct output. If the initial standardization were made at pH 10 and the slope adjustment at pH 13.0, the reading displayed on placing the electrodes again in pH 10 buffer would not be 10. As shown in Figure 2.5, the reading would correspond to that of point B. To avoid this interaction between standardization and slope adjustment controls, it is necessary to set the standardization point to zero potential.

The *zero* (isopotential) control provides the flexibility to standardize at a point other than pH 7 and then make a slope adjustment without affecting the standardization point. It provides greater accuracy by allowing the two-point calibration to take place over a narrower pH range. The zero control provides a potential to offset the ideal standardization potential and thus provide zero millivolts at a point other than pH 7. The zero control is first adjusted to pH 10.0 when in the standby mode which separates the electrodes from the meter. Then the electrodes are standardized in a pH 10.0

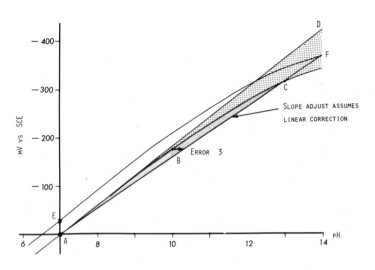

Figure 2.5
Wide-Range Slope Correction

			Calibrated (pH)	
Error Due to		Zero	Standard	Slope
▨ 1	Nonlinearity of electrode response	7	7	13
☐ 2	After slope adjustment			
3	Slope adjustment, changes in standard potential	7	10	13

$\overline{AD}$ Ideal Nernst response

Electrode response

$\overline{EF}$ Before standardization
$\overline{ABC}$ After standardization

buffer solution. The electrodes are immersed in a pH 13.0 buffer solution and the display is made to read 13.0 using the slope control as illustrated in Figure 2.6, without affecting the initial standardization point at pH 10.0. The zero control is used in conjunction with the slope control to provide a narrower range (10–13) over which calibration can be performed.

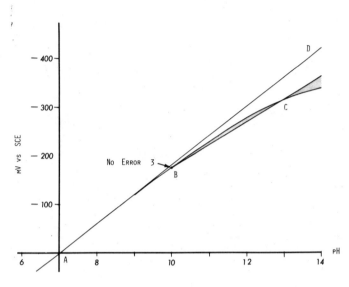

Figure 2.6
Narrow-Range Slope Correction

Calibration (pH)

Zero	Standard	Slope
10	10	13

Calibrated for pH 10 to 13 with less error 2

2.4.1 Automatic Standardization

On some manufactured pH meters the standardization adjustment can be done automatically. This is a convenience feature when frequent standardizations are done to a single buffer point. This is accomplished by matching the electrode output to a buffer value which is set to zero millivolts with the zero control. For example, if frequent standardizations with pH 7.41 buffer solution are being performed, the zero control is adjusted until the display indicates 7.41 when in stand-by. When the electrodes are immersed in pH 7.41 buffer solution and the autostandardization mode is initiated, the meter searches for a zero potential by offsetting the electrode potential. Since zero potential is also the standardization point, the meter is now standardized. If the meter is standardized, followed by a rinse, followed by sample measurement, and then returned to the buffer solution for restandardization, and this sequence is followed for a great number of samples, the autostandardization feature is very useful.

Whether manual or automatic standardization is employed, the criteria of this control are resolution, range, and stability. Often a meter which reads to 0.001 pH has two controls, coarse and fine, to provide the needed resolution. The range is typically ±100 to 200 mV (>1-3 pH units) to offset aging electrodes with a high asymmetry potential or to standardize an ion-selective electrode to a particular value. The stability is indicated by the amplifier drift specification which is typically less than a millivolt in 12 hours.

2.5   OTHER FUNCTIONS

Although the millivolt function of a pH meter is mostly used with electrodes other than pH glass electrodes, it is an integral part of the meter and should be discussed briefly. The major difference between the pH and millivolt functions is the temperature compensator. It is active in the pH function and inactive in the millivolt function. All potentimetric electrodes follow the Nernst relationship, and the slope varies with the number of electrons (n) involved in the reaction. In the pH function, the temperature compensator incorporates an "n" value of one. Since other electrodes or reactions may involve other "n" values, it becomes difficult to make a universal temperature compensator. Thus the millivolt function is not affected by this control, although the electrodes and the sample are affected by temperature as with pH measurements. Therefore, when stating a millivolt value, it is stated as millivolt versus the reference electrode at a specific temperature.

Some meters have a millivolt and an absolute millivolt function. The function labeled "mV" is a relative reading; that is, it is relative to the standardization control potential setting or millivolt values obtained from standard solutions. In the absolute millivolt mode, the standardization control is inoperative and the potential reading of a sample is relative to zero millivolts. The input will read zero when the meter is placed in stand-by.

2.6   TYPES OF METERS

There are several basic types of pH meters which generally fall into categories of low cost, utility, and research. The low-cost-type meter is used more as a screening tool, often for field application, with readability of approximately 0.1 pH unit. The utility-type meter has good accuracy, ease of operation, and is generally used in the quality control laboratory. Its readability is generally about 0.05 pH unit for meters with standard scale and about 0.007 pH unit for meters with expanded scale capabilities. The research category provides the accuracy required to observe deviations less than a hundreth of a pH unit. The readability will be about 0.002 or better and requires a good deal of knowledge about other parameters that influence the pH measurement in order to take advantage of this readability.

Selection of a pH meter should be based on the application, accuracy, and reliability required. The application dictates the power requirements (portable or line operated), the readout, and simplicity of operation. For example, if the pH meter is to be used in both the laboratory and in the field, perhaps a meter with a nickel–cadmium rechargeable battery would be preferred. An application which involves reading numerous measurements may sway preference to a digital readout for use with a printer. The ease of operation, such as handling the electrode stand and support, viewing angle, and pushbutton operation, would also be a factor in selection for this type of application. Secondary applications may require other features, such as a log scale for ion selective measurements or polarizing current for titrations. Ruggedness as well as reliability are often requirements for industry. A built-in test function adds an element of reliability since it immediately pinpoints whether a problem exists with the meter or the electrodes.

2.7   PERFORMANCE SPECIFICATIONS

Manufacturers of pH meters may publish specifications such as those listed in Table 2.4. Some specifications may be misleading, because the conditions for which they are valid are not always obvious. For example, a relative accuracy specification of ±0.07 pH may be more accurate than one stated as ±0.03 pH within 2 pH units of standardization. In the first case, the manufacturer may include the entire meter scale linearity, while in the second case only a small portion of the scale is considered. A specification on a temperature compensating control is usually better between 10° and 60°C than it is outside this range since the extremes of this potentiometer are not as accurate.

In order to compare pH meters from different manufacturers or in order to verify performance specifications, it may become necessary to check the calibration of the meter. The simple troubleshooting procedures to verify gross malfunctions are described in Chapter 7. For detailed inspection, however, the following procedures apply mainly to meters with 0.02 pH readability or better.

2.7.1   Drift

The purpose of the drift test is to detect amplifier drift that causes errors in long-term measurements. A recorder is used to make a continuous recording overnight or at least for 5 hours. In order to make this test valid, a high impedance should be put on the input. Many manufacturers sell a test resistor which is shielded and plugs into the glass and reference inputs. Therefore, the steps involved are:

(a)   Short the glass and reference inputs using a test resistor.
(b)   Place the pH meter in the millivolt mode.
(c)   Then connect the pH meter to a potentiometric recorder and adjust the recorder output to 100-mV full scale.
(d)   With the readout activated, adjust the standardization control until the readout displays about 50 mV.

TABLE 2.4

pH Meter Specifications

	0.01 Digital meters			Analog expanded meters		0.001 Digital meters	
pH range	0.00–13.99 pH	0.00–14.00 pH	0–14 pH	0–14 pH Any 1.4 pH units	0–14 pH Any 3 pH units	0–14.000 pH	0–14 pH
mV range	±1999 mV	±1999 mV	±1800 mV	0–±1400 mV	0–±1400 mV	±1999.9 mV	0–±1800 mV
Relative accuracy	±0.01 pH ±1 mV or 0.1%	±0.01 pH	±0.02 pH ±1 mV	Any 140 mV ±0.07 pH ±0.007 pH exp ±7 mV ±0.7 mV exp	0–±300 exp ±0.05 within 4 pH units ±0.01 pH exp	±0.001 pH at 30°C ±0.01 mV	±0.002 pH ±0.02 mV
Repeatability	—	±0.01 pH ±1 mV	±0.01 pH ±1 mV	±0.02 pH ±0.002 pH exp ±2 mV ±0.2 mV exp	±0.02 pH ±0.005 pH exp	±0.1 mV ±0.001 pH	±0.001 pH ±0.1 mV
Slope	80–100%	80–105%	80–100%	80–105%		80–105%	80–100%
Temperature compensation	0–100°C ±1°C, 10–60°C ±2°C, 0–10 and 60–100°C	0–100°C	0–100°C	0–100°C ±1°C, 10–60°C ±2°C, 0–10 and 60–100°C	0–100°C	0–100°C ±1°C, 10–60°C ±2°C, 0–10 and 60–100°C	0–100°C
Input impedance	10^{13} ohms	>10^{13} ohms	>10^{13} ohms	>10^{13} ohms			
Bias current	<5 picoamps	<1 picoamp	<30 picoamps	1 picoamp		<0.1 picoamp	
Drift	<100 mV/°C	<3 mV/10 hours—		<0.5 mV/12 hours	<0.005 pH/day	<0.1 mV/12 hours	
Zero range	6–8	4–10		4–10		6–8.000	
Standardization control	—	±3.5 pH	±220 mV (3.5 pH)	Coarse, 0–14 steps Fine, ±3 pH units		±220 mV (±3.5 pH)	3.5 pH at 25°C

34

(e) Allow the recorder to run for about 10 hours and note the millivolt drift during this time period. A drift of less than 2 mV should be observed for most accurate meters.

Testing Box

An accurate millivolt source which can provide up to ±520 mV is needed for verification of the remaining meter specifications. Once again, the degree of verification is much greater if the input incorporates a high resistance similar to that of an electrode. At the resistance values of 10^8 to 10^9 ohms, shielding of the input is required in order to maximize stability. Thus a shielded millivolt source with high resistance is needed. This source placed in a metal box would have a circuit as shown in Figure 2.7. It is also available from many manufacturers of pH meters, since it would be useful in servicing any meter.

2.7.2 Response

If the operational amplifier used in the meter (FET, field effect transistor) is not of good quality or if the amplifier becomes saturated as the result of a faulty standby switch, a slow response measurement could result. Slow response, however, is much more likely to be the result of the electrodes or technique as described in Chapter 5. The steps to test a pH meter response include connecting a 500-mV source to the input and then noting the time required between a standby and an activated readout.

(a) Connect a 1000-megohm test resistor to the glass input and connect one lead from the millivolt source to this resistor.
(b) Connect the other lead of the millivolt source to the reference input and set 500 mV on the millivolt source.
(c) Activate the readout and adjust the standardization control until the display reads 500 in the millivolt mode.
(d) It should take less than 2 seconds for the pH meter to reach 98% of the final reading when the readout is taken from the deactivated to activated position.

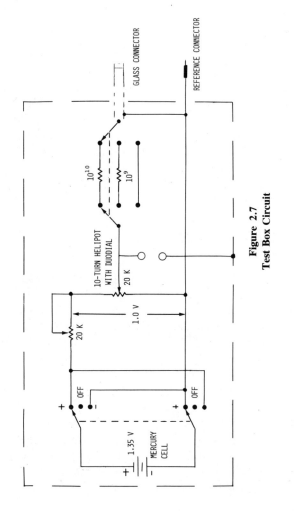

Figure 2.7
Test Box Circuit

2.7.3 Temperature Linearity

Most meters are calibrated at 30°C and the temperature potentiometer linearity is depended upon to maintain accuracy. Before
checking the relative pH accuracy, the temperature compensator
should be calibrated. In other words, does 30°C on the temperature
scale markings provide the correct slope? Since the temperature
compensator has the greatest effect at the extremes of the pH scale
as previously described, its calibration should be checked at either
extreme. The test consists of applying a millivolt source equivalent
to pH 14 at various temperatures. These values are listed in Table
2.5.
The steps to accomplish this test are:

(a) With zero millivolts applied to the pH meter inputs, adjust
     the standardization control until pH 7.00 is displayed.
(b) Apply the millivolt values listed in Table 2.5 and adjust the
     temperature compensating control until pH 14 is displayed.
(c) Note the variation between this temperature and the tem-
     perature listed in Table 2.5.

Since most pH measurements are made within the 10° to 60°C
range, the manufacturers often ensure that the greatest accuracy is
in this range. The average pH unit represented by 1°C is 0.003 per
pH unit from standardization as explained earlier. With this in
mind, ±1°C between 10° and 60°C should provide sufficient accuracy for a meter whose readout is to ±0.01.

TABLE 2.5

pH 14 mV Values

Temperature (°C)	Millivolts
0	−379.4
30	−421.0
60	−462.7
100	−518.3

2.7.4 Relative pH Accuracy

The word "relative" applies to the accuracy of the standard
buffer solution, the proper working electrodes, and the use of
proper technique. Meter accuracy alone is not a significant specification since an entire system is involved in the measurement. The
accuracy of entire system, however, depends on the many options
available to the users, and the specifications which the manufacturer states are often only for the meter. The meter accuracy can
be tested to obtain its magnitude and verify that it is not adding
significant error. The steps involved in this test are:

(a) Set the temperature compensator at 30°C or the temperature value determined to be the equivalent of 30°C by the
temperature linearity test.
(b) With zero millivolts applied to the pH meter inputs, adjust
the standardization control until pH 7.00 is displayed.
(c) Apply the millivolt values listed in Table 2.6 from a
millivolt source.
(d) Note the pH value at each of the millivolt values and
calculate the difference between the observed and the listed
pH values.

How does the relative pH value compare to the stated pH
accuracy? The relative pH accuracy should be at least 1/5 of the
readability of the pH meter.

2.7.5 Bias Current

This specification is an indication of the ability of the meter to
handle high resistance samples or electrodes, without significant
noise or error. The test consists of comparing identical millivolt
inputs with and without a high resistance in the circuit. Proceed as
follows:

(a) Connect a properly shielded millivolt source to the inputs
(glass and reference) of a pH meter.

TABLE 2.6

Millivolt Values for pH Integer Values at 30°C

pH		Millivolts
+	−	
6, 8		± 60.15
5, 9		±120.3
4, 10		±180.5
3, 11		±240.6
2, 12		±300.8
1, 13		±360.9
0, 14		±421.0

(b) After calibrating the pH meter to zero millivolts, apply 100 mV from the source to the pH meter and note the readout.

(c) Then connect a 1000-megohm test resistor to the glass input that is connected to the millivolt source. If the previously described test box is used, this resistance can be switched into the circuit.

(d) Apply the same 100 mV and note the readout. There should be less than ±1.0 mV difference between the two readouts for a meter with 1 picoamp bias current.

$$E = IR$$

$$= 10^{-12}\,\text{amp} \times 10^9\,\text{ohms}$$

$$= 1 \quad \text{mV}$$

Since the display uncertainty enters into this reading, the millivolt value would be slightly greater than one.

These tests are suggested only when verification of performance specifications is necessary to achieve a high degree of accuracy. It is much more vital that the performance of the electrodes be tested since they are more likely to be the source of problems.

Chapter 3

Electrodes

This chapter is concerned with electrodes. Section 3.1 provides information about the structure and characteristics of glass electrodes along with possible sources of error, such as sodium ion interference. This should assist in optimum selection and use of electrodes. In Section 3.2, the structure and functioning of reference electrodes are discussed. The liquid junction between the reference electrode electrolyte filling solution and the sample can introduce a junction potential, which is a major source of error in pH measurement. This subject is discussed in detail. The combination electrode is compared with an electrode pair and its advantages in small-volume samples or flat-surface measurements are discussed in Section 3.3.

3.1 GLASS ELECTRODES

The glass electrode must meet a great number of demands. It must respond with good span or near ideal millivolts per pH unit throughout the pH scale with little error in very alkaline or acid solutions. It is desirable to have the glass at low melting point for fabricating. It should be chemically durable to act as a pH electrode and physically durable for practicality. Yet the glass cannot be too thick since a relatively low resistance is desirable. The electrode must have high hygroscropicity since a hydrated layer is necessary for the glass bulb to have pH response. It should also have a low asymmetry potential.

All of these requirements limit the glass electrode composition and configuration and make glass electrode manufacturing an art.

3.1.1 Construction

The glass electrode is constructed with pH sensitive glass in the form of a dome, bulb, or somewhat flat surface at the immersing tip. On the inside are an internal reference electrode element and filling solution. The potential of the glass electrode and its change with temperature are partially determined by these two components. Since the internal solution and internal reference element are unchanging, their major function is to establish stable potentials while completing the circuit.

The internal filling solution is a buffered chloride solution of constant hydrogen ion and chloride ion concentration. The hydrogen ion establishes an ionic exchange equilibrium with the inner glass surface to produce the phase-boundary potential (E_3, Figure 1-3), while the chloride establishes a constant potential with a silver–silver chloride internal reference element (E_1, Figure 1-3) (also see Figure 3.1).

The potential at the external glass surface is developed as a result of ion-exchange reactions with the solution in which it is immersed. The glass structure must maintain anionic sites for the ion exchange. Silicon dioxide of greater than 50% by composition provides this characteristic. The stability, electrical conductivity, and sodium errors of an electrode are somewhat dependent on the ionic properties of other elements (modifier elements) in the glass. The ease with which ionic transfer between glass and solution can occur is the result of these components.

The glass composition usually consists of an alkali metal(s), a bivalent or trivalent metal(s), and silicon dioxide in relatively the ranges shown in Table 3.1. The composition is varied to produce

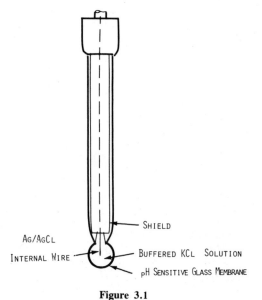

Figure 3.1
Glass Electrode Components

the desirable compromise between response and resistance. Some manufacturers of glass electrodes produce several types of electrodes, those with narrower pH range (e.g., 0–11) and those with full pH range (0–14). Electrodes with narrower range usually have lower resistance than full-range electrodes. Thus, they may provide better response in low-temperature, nonaqueous and any other

TABLE 3.1

Typical Glass Compositions

Electrode pH range	Alkali metal(s) 17–32%		Rare earth or alkali earth metal(s) 3–16%		Silicon dioxide 60–75%
	Li_2O	Cs_2O	BaO	La_2O_3	SiO_2
0–14	28	2	5	2	63
0–11	25		7		68
0–14	28	3		4	65

sample whose pH value is within the useful range or when the sodium ion concentration is low enough so that sodium ion error is negligible.

The standard laboratory glass electrode is about 12 mm in diameter by about 12 cm long. Glass electrodes, however, are available in a variety of sizes and shapes ranging from microcellular or one-drop size to large-process size electrodes.

3.1.2 Glass Bulb Function

The ability of an electrode to respond to changes in pH is usually associated with the water content of the glass. The surface of the glass bulb swells slightly when immersed in a solution, and a hydrated layer is formed as the water penetrates into the silicate network. This layer seems to facilitate the movement of ions in the glass and lower the electrical resistance. In a simplified description, a proton transfer occurs at the phase boundary between the glass and the hydrated layer. An ion-exchange equilibrium is established between the hydrogen ions in the solution and the alkali metal ions in the glass. The potential of this equilibrium is dependent upon the hydrogen ion activity. One reason the hydrogen ion has the ability to pass the phase boundary more easily than other positive ions and penetrate to the silicon–oxygen network is its small size.

The alkali metal ions within the glass move slightly toward one surface or the other giving rise to an electrical potential. The direction in which these ions move depends on the potential. The different components are shown in Figure 3.2, where the letters represent the following items:

A   glass membrane,
B   hydrated layers,
C   phase boundary potentials,
D   diffusion potential,
E   internal element,
F   internal filling solution—a pH buffered chloride solution.

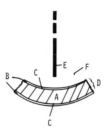

Figure 3.2
Glass Membranes

The phase boundary potential (E_p) is controlled by the ion-exchange equilibrium between the ions in solution and those immediately inside the glass surface. The diffusion potential (E_d) is normally constant and is the result of interdiffusion of ions in the glass. It does affect the selectivity and pH range of the electrode.

The asymmetry potential (E_a) is the difference in potential between the inner and outer surfaces with the same solution inside and out.

The internal element is normally a silver–silver chloride element which establishes a potential (E_i) with the pH buffered chloride filling solution.

Thus the glass electrode potential (E_g) is the result of several potentials:

$$E_g = E_p + E_d + E_a + E_i$$

If the electrode is allowed to dry, it may partially lose its pH function. Therefore it is imperative that a glass electrode which is used for nonaqueous pH measurements be soaked periodically in water to rejuvenate it. Even with partial dehydration, glass electrodes function properly for a moderate time period in nonaqueous solutions having a dielectric constant as low as 2.3 (see Table A.2).

3.1.3 Electrode Properties

As the alkaline constituents are extracted from the glass by water, the silicon–oxygen network is decomposed and limits the electrode *life*. This occurs from both sides until the electrode fails. A thicker electrode would last longer, but as thickness is increased the electrical resistance is increased. There has to be a compromise between chemical and electrical resistance. With proper care and storage, a glass electrode should last more than 2 years.

A. Resistance

The resistance of glass electrodes depends on the glass composition, the hydrated conditions of the bulb, the bulb configuration or thickness, and the temperature.

The resistance of the membrane does not affect the pH response. If the glass is too thick, however, the electrode may depart from normal response. With a high-resistance glass membrane, a noisier reading and a pH error are more likely. As the combined resistance of the pH system, of which the glass electrode component is normally the greatest, approaches the input resistance of the pH meter, the noise and pH error increase. This is further discussed in Section 2.1.

B. Temperature

Another parameter which can cause higher resistance values is temperature. The resistance of the glass membrane approximately doubles for every 7°C decrease in temperature. In other words, a glass bulb with a resistance of 100 megohms at 25°C has a resistance of about 800 megohms at 4°C. Therefore, it is more important to use a glass electrode with low resistance if low temperature measurements are anticipated. Generally this means a glass bulb with limited pH range (0–11).

C. Shielding

With the high resistance values encountered in glass membranes comes the requirement for shielding. In some manufactured electrodes, the shielding is provided by a metallic label which surrounds the internal element of the glass electrode for the entire length of the glass body. Normally, the glass bulb begins just below the end of the shield. Therefore, the glass electrode should be immersed in solution to at least the level where the shielding ends. The solution itself provides the shielding for the glass bulb. If a gap between the end of the shielding and the solution exists, a noisier reading will be observed than with a completed shield (see Figure 3.1).

On a combination electrode the shielding is provided by the reference electrode filling solution. The metallic shielding on a combination may extend only partially down the electrode body and rely on the filling solution to complete the shielding. Thus, the filling solution level should be kept sufficiently high to complete the glass electrode shielding.

3.1.4 Asymmetry Potential

If both sides of the glass membrane are in contact with solutions of identical pH but develop potentials which are different, the condition is called the asymmetry potential. The potential difference is normally a few millivolts. Any influence which alters the composition and ion-exchange properties of the glass can cause asymmetry. For example, this can be caused by absorption of protein which disturbs the exchange capacity of the electrode, or by dehydrating the surface layer by using it in nonaqueous solutions. The magnitude of this potential varies with the pH of the solution in contact with the glass, with age, with temperature, with membrane configuration, and with time.

The asymmetry potential term has also been used to define the potential differences between electrode pairs when placed in identical solutions. Although the electrode asymmetry potential may change slightly on day-to-day use, it is compensated through use of the standardization control of the pH meter. It is only when the rate of change becomes a source of error in long-term measurements or the magnitude exceeds the capacity of the standardization control to compensate that this potential becomes a problem. Either electrode rejuvenation or replacement becomes necessary at this point. Also, the asymmetry potential of the glass electrode affects the isopotential point. If the glass electrode is stored in a soaking solution so that the glass membrane does not dehydrate, the asymmetry potential is likely to remain constant.

3.1.5   Span

The main criterion for judging the performance of a glass electrode is the span or "electromotive efficiency." In other words, how many millivolts per pH unit does the electrode produce and how close does this come to the ideal output? In the range of pH 1 to 10, close to 99% ideal output should be observed (i.e., 59.16 mV/pH unit at 25°C). Therefore, an electrode standardized at pH 4.01 buffer solution and then immersed in pH 9.18 buffer should exhibit the correct pH value within 0.05 pH unit.

The response of the glass electrode, however, is impaired in strong alkaline and acidic solutions. The error in alkaline solutions requires a positive correction (add to the observed pH value), whereas the acidic correction is negative.

3.1.6   Sodium Ion Error (Alkaline Error)

The sodium ion error is the result of sodium ions penetrating into the silicon oxygen network and causing a potential response. It is called the sodium ion error even though other small cations can also cause error. However, other ions, such as lithium and potassium, usually cause significantly less error and are not encountered as frequently in high concentration at high pH values. The effect of this potential can be related to the predicted potential by adding a term to the Nernst equation (10):

$$E = E^0 + S \log(a_{H^+} + Ka_{Na^+})$$

The value of K, the selectivity coefficient, relates the ion-exchange equilibrium of the sodium ion compared to that of the hydrogen ion. It is a temperature-dependent constant in the law of mass action for ion-exchange equilibrium between sodium and hydrogen in the glass and in solution.

Manufacturers may express the value of this error in different ways. The essence of the correction is to compensate for response to the sodium ions by adding some value to the observed reading (see Table 3.2).

TABLE 3.2

Correction for Sodium Ion Error

25°C	Glass type	
	0–11	0–14
0.1 *M* Na⁺		
pH 11	0.05	—
12	0.15	0.01
13	0.45	0.03
14	1	0.1
1.0 *M* Na⁺		
pH 11	0.15	0.01
12	0.4	0.03
13	1	0.1
14		0.3

If cations other than hydrogen ions participate in the ion-exchange process, they may contribute to the electrode potential and thus give rise to the measuring error.

3.1.7 Acid Error

The acid error is different from the alkaline error in that it changes very little with temperature and is a negative error where a greater number of millivolts per pH unit is observed. With high acid activity, the water activity is reduced and some reduction in the hydrated layer occurs. The magnitude of this error may vary between 0.1 pH in a 2 *M* acid solution to more than 8 pH units in concentrated acids. The magnitude of this error depends on pH, temperature, exposure time, and size of the anions in the test solution.

3.1.8 Care and Cleaning

The main concern for proper glass electrode care is to keep the glass bulb soaking in solution (see Section 3.5). If the electrode exhibits one of the following, rejuvenation may be necessary: a lag in response, undue sensitivity to physical movement of the electrode, a short span between two buffers, or abnormal potentials—thus an inability to calibrate the pH meter with the standardization control.

The process of rejuvenation may depend on the electrode use. For example, if the electrode has been used in an organic compound which may have coated the bulb, then the solvent that would dissolve the compound would be the first rinse. This should be followed by a rinse which would remove the previous rinsing solution but be more polar, for example, an alcohol.

A more universal rejuvenation consists of an acid–base–acid cycle; that is, to immerse the bulb in $0.1\,M$ HCl followed by immersion in $0.1\,M$ NaOH and then $0.1\,M$ HCl, each for a 5-minute period (see Figure 3.3).

If this cycling fails to rejuvenate the glass electrode, a new glass surface can be generated by etching its surface. The bulb is dipped in a 20% ammonium bifluoride solution for 10 to 30 seconds and then rinsed thoroughly with water. Then the electrode is immersed in $5\,M$ hydrochloric acid for a few seconds to remove the fluorides and rinsed again with water.

This rejuvenating procedure will reduce the electrode life by making the glass bulb thinner or cracking the glass. Therefore, it should be used only after other methods of rejuvenation have failed to restore the electrode span or response.

0.1 — 1 M HCL   0.1 — 1 M NaOH

Figure 3.3
Rejuvenation

3.1.9   Performance Specifications

Some performance specifications such as span, response, and repeatability can be tested. Only a few manufacturers actually state these specifications, but some typical values are listed here:

Span error   ±0.02 pH/pH unit from standardization,
Response   10 seconds to 98% of final reading,
Repeatability   ±0.02 pH for a series of 10 measurements.

A.   *Span*

The span specification will vary with the type of glass and its configuration. Small bulbs or higher resistant glass are more likely to have shortened span. If most pH measurement values are near the standardization point, span becomes less important and an electrode with a slightly short span has little effect on the accuracy. If, on the other hand, the measurement values are over a wide pH range, an electrode with short span has more effect on the accuracy.

The span error for most electrodes should fall within ±0.02 pH per pH unit from standardization. This span specification is usually exceeded if proper care and technique are employed. With increased span error, through aging or use, a minimum acceptable error should be selected based on the application. A suggested span error value of ±0.06 pH per pH from standardization may be selected as a point of which rejuvenation or electrode replacement may be necessary. This would mean that a deviation of less than 0.3 pH should be observed after standardizing the electrodes in pH 4.01 NBS buffer and then measuring pH 9.18 NBS buffer. The second buffer 5 units from standardization times 0.06 pH unit yields the 0.3 possible deviation (see Figure 3.4). If the electrode fails the span test, that is, gives a value below pH 8.9, and does not improve with rejuvenation, the glass bulb may be scratched. Examine the bulb under a microscope to determine if this is the case.

The span specification is the main criterion for judging glass electrode performance. Unfortunately, this test also assumes a

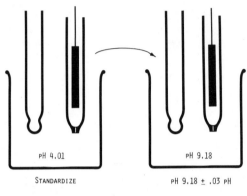

PH 4.01   PH 9.18

STANDARDIZE   PH 9.18 ± .03 PH

Figure 3.4
Span Test

properly functioning reference electrode and pH meter. Therefore, a negative result may not be the result of a malfunctioning glass electrode.

B. Response

A glass electrode response test can use the same buffer solutions as employed for the span test. After standardization of the pH meter in pH 9.18 buffer, the reference electrode tip is immersed in pH 4.01 buffer to preequilibrate for a period of 5 minutes. This eliminates any response time due to the reference electrode. After this time period, rinse the glass electrode with pH 4.01 buffer and immerse the bulb in the same buffer solution with the reference electrode. Record the pH value versus time or observe the reading after 10 seconds. The reading after the 10-second period should be 98% of the final reading; that is, the meter should read 4.11 or less within the 10 seconds. If the electrode fails this test, rejuvenation may help to increase its response. Response time for electrodes is discussed in detail in Section 5.3.

C. Repeatability

The repeatability test for glass electrodes is a series of measurements in the same solution with rinsing between each measurement. A series of 10 measurements, in pH 4.01 buffer with a distilled water rinse and a tissue blot between each measurement, should agree within ±0.02 pH unit. Be sure to blot and not wipe the electrode as cautioned in Section 5.1.

3.2 REFERENCE ELECTRODES

The purpose of the reference electrode is to provide the second electrode of an electrical cell whose potential is measured for the determination of pH. It must have a stable and reproducible potential to which the glass electrode potential may be referenced. The reference electrode completes the circuit by contacting the sample solution through a liquid junction. It is this liquid junction that is most often a problem in pH measurement and will be discussed in detail in Section 3.2.3.

3.2.1 Construction

The reference electrode incorporates an internal element, normally calomel or silver–silver chloride, and an electrolyte filling solution contained in a glass or plastic body salt bridge which surrounds this element and terminates in the liquid junction. The various parts of a reference electrode are shown in Figure 3.5.

The internal element appears as a grayish-white cylindrical pack with shiny mercury at the top of the element, if it is a calomel internal. This mercury–mercurous chloride half cell provides a potential of 244 mV versus the normal hydrogen electrode at 25°C if it is surrounded by saturated potassium chloride filling solution. It is important that this element be kept wet and uncontaminated in order to provide a stable and reproducible potential. With use, the pack may show some separation within the element tube, but this usually does not cause error or deviation of its potential.

A standard silver–silver chloride reference electrode provides 199 mV versus the normal hydrogen electrode if it is surrounded by a filling solution saturated with both potassium chloride and silver

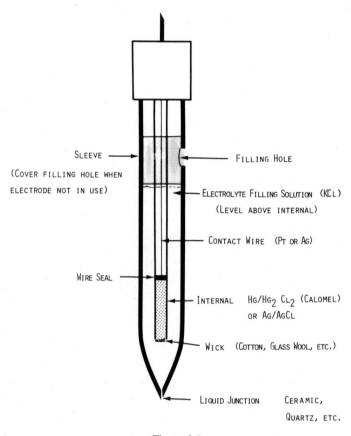

Sleeve ─────→
(Cover filling hole when
electrode not in use)

Wire Seal ────────→

Filling Hole

Electrolyte Filling Solution (KCl)
(Level above internal)

Contact Wire (Pt or Ag)

Internal   Hg/Hg$_2$ Cl$_2$ (Calomel)
               or Ag/AgCl

Wick (Cotton, Glass Wool, etc.)

Liquid Junction   Ceramic,
                        Quartz, etc.

Figure 3.5
Reference Electrode

chloride. Either internal element will provide a stable and reproducible potential for most measurements. The silver–silver chloride internal element is preferred when making high-temperature pH measurements ($>50°C$) since calomel may disproportionate above this temperature.

Since it is important to keep the internal element wet and surrounded by a filling solution of known concentration, manufacturers of reference electrodes ship the electrode filled with filling solution. In order to prevent leakage of the filling solution through the filling hole during transit, the hole is often sealed with tape. The tape must be removed and the rubber sleeve which covers this hole must be moved down to uncover the hole when the reference electrode is in use. If this is not done, as the filling solution leaks through the junction, a vacuum is created inside the electrode until the filling solution no longer flows. A drifting or unstable reading may result.

The reference electrode is also sold with a cap containing solution over the junction. This is to keep the junction wet and permeable at all times. It is important that the junction should be kept wet in order to function properly. (See Section 3.5.)

3.2.2   Junctions

The liquid junction allows leakage of the filling solution into the sample. Manufacturers of reference electrodes make the junctions of various materials and with various configurations. The most common materials are asbestos or quartz fibers, ceramic plug carborundum frit, and ground-glass sleeve construction (see Figure 3.6). There is a variety of junctions since no one junction has been found to have universal application. The main differences are the junction potential and the rate at which the filling solution leaks through the junction. Therefore, the choice of liquid junction type may depend on the application. For example, an annular ceramic provides low junction potential for most samples but does not perform as well as a quartz junction in strong acid solutions.

Probably the closest junction to being universal is the sleeve junction. It provides a low junction potential for most samples, is easily cleaned, and is recommended when it can be applied. The disadvantage of this type of junction is the high flow rate of filling solution. This requires frequent filling of the electrode with filling solution. The high flow rate (>1 ml/day) also presents a significant contamination problem for small-volume samples. The flow rate of the sleeve junction can be reduced by applying a small amount of silicone grease to the top of the glass body taper. Then as the sleeve is put into place and rotated, the grease provides a thin film seal and limits the filling solution flow. This junction is cleaned far

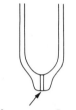

CERAMIC AND FRIT

QUARTZ AND ASBESTOS

SLEEVE

ANNULAR

Figure 3.6
Junctions

easier than any other type of junction by simply removing the sleeve.

The quartz junction has a slow flow rate and provides good stability in nonaqueous, low ionic strength, and strong acid samples. Table 3.3 provides an indication of performance for different junctions in various types of solutions.

The rate at which the filling solution leaks from the junction will vary with the type of junction material. Normally a junction with a moderate flow rate will provide low resistance and a low junction potential. The slower flowing junctions provide less contamination but are more easily clogged than a fast flowing junction. Typical junction flow rates are shown in Table 3.4.

The extent of sample contamination by the filling solution, of course, also depends on the length of time the reference electrode is in contact with the sample and the volume of the sample. One method of reducing contamination is by use of more dilute filling solutions. This may increase liquid junction potential. It may also

TABLE 3.3

Reference Junction Performance

Solution	Annular ceramic	Quartz	Asbestos fiber	Sleeve
Oxidizing and reducing agent	Satisfactory	Recommended	Recommended	Satisfactory
Precipitates and colloidal suspension	Satisfactory	Satisfactory	Satisfactory	Recommended
Strong base	Recommended	Satisfactory	—	Satisfactory
Strong acid	—	Recommended	Satisfactory	—
Low ionic strength	Recommended	Recommended	Recommended	Satisfactory
Nonaqueous solutions	Satisfactory	Recommended	—	Satisfactory

require establishing a new reference potential as suggested by Table 3.5.

It may be necessary to eliminate the leakage of the potassium chloride filling solution from the junction into the sample (for example, a sample with silver ion). An auxiliary salt bridge, which is a glass body with a junction at one end, is available from some electrode manufacturers. This can be filled with a salt solution which does not contain a contaminant. The reference electrode is then placed in the salt bridge to make contact with the salt solution (see Figure 3.9). Reference electrodes which have this double junction feature built in are also available.

Contamination of the internal element with by the sample can result if the filling solution level is not maintained. The positive flow of filling solution from the electrode into the sample is

TABLE 3.4

Junction Flow Rates

Ceramic plug	≈ 5 μmoles Cl^-/hour
Quartz fiber	≈ 50 μmoles Cl^-/hour
Annular ceramic	≈ 50 mmoles Cl^-/hour
Sleeve	≈ 500 mmoles Cl^-/hour

dependent on the head pressure; that is, the filling solution level is higher than the sample level. If the filling solution level is lower than the sample level, the flow can reverse and the sample can get into the filling solution, which can cause an unstable potential. It is desirable to maintain a high level of filling solution to provide more head pressure. Reference electrodes which have a side arm for pressurization are available. This provides greater assurance of a free-flowing junction and is most often used when long term measurements are encountered (see Figure 3.7).

Most filling solutions are saturated or nearly saturated with a salt (see Section 3.2.4). Because of this, crystals normally settle at the bottom of the electrode, which does not interfere with the electrode performance. In fact, the crystals assure saturation even if the temperature of the filling solution changes slightly. This concentration of salt also causes the encrusting of salt around the filling hole if the sleeve does not cover the hole.

Over a period of time, however, the crystals at the bottom of the electrode may become packed and hinder the flow of filling solution through the junction. The crystal pack can be eliminated by shaking the filling solution out through the filling hole and filling

TABLE 3.5

Reference Standard Potentials

| Temperature (°C) | Calomel | | | | AgCl | | |
	Saturated	3.5 N	1 N	0.1 N	E^0	$E^0 + E_j$ 3.5 M	Saturated KCl
0	260.2				236.6		
10	254.1	255.6			231.4	215.2	213.8
20	247.7	252.0		335.8	225.6	208.2	204.0
25	244.5	250.1	283.	335.6		204.6	198.9
30	241.1	248.1		335.4	219.0	200.9	193.9
40	234.3	243.9			212.1	193.3	183.5
50	227.2				204.5		
60	219.9				196.5		
90	—				169.5		

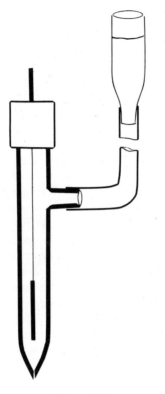

Figure 3.7
Side Arm Reference

Reservoir with filling solution creates additional head pressure.

the electrode with distilled water. Then allow the KCl crystals to dissolve before replacing the water with normal filling solution.

3.2.3 Liquid Junction Potential

When a pair of electrodes is placed in a solution to make a potentiometric measurement, many potentials are present (Figure 1.3). Most of these, however, are stable and therefore not a source of problems.

A potential at the junction between the reference electrode filling solution and the sample (Figure 1.3, E_7) is present any time these two solutions are different. The potential results from interdiffusion of ions in the two solutions. Since ions diffuse at different rates, the electrical charge will be carried unequally across the junction. This results in a potential whose magnitude and stability depends on the composition of the solutions and the type of junction.

This section attempts to show how to reduce and/or stabilize this potential in order to decrease this error.

Difficulties encountered with electrode measurements are most often traced directly to the reference electrode junction. Since this portion of the electrode is of such great importance to the measurement, its selection should be considered carefully. A pH meter with readability of at least ±0.01 pH is usually employed for application when a high degree of accuracy is required. If the wrong reference electrode is used, an error or drift may occur that is of such magnitude that it could be observed even on a meter which has a readability of ±0.1 pH. Some examples of typical problems arising from liquid junction potentials are shown in Figure 3.8.

The top example shows a recording of pH with time under static and stirred conditions with two different values observed. The next example shows a variation of pH reading noise with varied stirring rates. The third example illustrates the long time that may be required to obtain a stable reading when a large or changing liquid junction potential exists. The last example represents drift or wandering of the reading due to this potential change. All of these example problems can be related to a liquid junction potential.

Assuming a free-flowing junction, there are basically two approaches to reducing or stabilizing the junction potential. The first approach is to change the junction material and/or configuration. Some junctions may provide excellent performance, while others provide poor performance in the same solution. The general discussion on junctions in Section 3.2.2 should act as a guide to reference junction selection for a particular application. The second approach deals with solvents and electrolyte composition of the filling solution which flows through the junction.

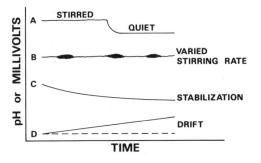

Figure 3.8
Problems Related to Liquid Junction Potentials

3.2.4 Filling Solutions

The reason potassium chloride is typically used as a filling solution is that it best meets all of the requirements for most applications. This filling solution, however, should not be the only one considered. In fact, in some applications this composition must be avoided, or perhaps another composition will provide better stability. This is particularly true when taking nonaqueous pH measurements. Thus the composition of the salt bridge may vary with the application. The conditions which a filling solution should meet are:

(1) The filling solution electrolyte should not react with the sample.
(2) The electrolyte should not contaminate the sample.
(3) The electrolyte should be soluble in the solvent used.
(4) The electrolyte should provide the dominant ions in the liquid junction.
(5) The diffusion rates of the anion and cations of the filling solution should be nearly equal.

An example of the first condition would be the reaction of the normal potassium chloride filling solution with a sample containing silver to form silver chloride. The second condition becomes more important with a fast-flowing junction when measuring a low concentration level over a long period of time. This condition is not as important in pH measurement as it is in ion-selective measurements in which the electrode may be more sensitive to the ions in the filling solution.

The third condition is a practical physical condition requiring the filling solution electrolyte to be soluble in the solvent of choice.

The last two conditions are required to establish a stable liquid junction potential. This is accomplished by providing a high concentration of electrolyte whose cation and anion components are balanced with respect to diffusion rates. The high concentration (3.5 N to saturation) of the electrolyte makes these ions dominant relative to sample ions. The ionic strength of the filling solution should be at least 10 times that of the sample and normally is much greater. This condition becomes obvious when high salt solutions, such as seawater, or a strong acid or base, are encountered and a large junction potential is observed. These types of samples may also require considerable time before a stable reading is established, depending on the type of junction and the sample. The condition for balanced diffusion rates provides equal transference of the positive and negative charges into the sample solution. The ability of an ion to carry a charge can be compared on the basis of ionic equivalent conductance (λ^0, mho-cm/equivalent/liter). Several cations and anions may be used in order to make the solution equitransferent. Examples of calculation if the electrolyte provides balanced transference follow:

$$(Z^+)(C^+)(\lambda_+{}^0) = (Z^-)(C^-)(\lambda_-{}^0)$$

where

Z is the charge,
C the concentration,
$^+$ the cation,
$^-$ the anion, and
λ^0 the limiting equivalent conductance.

Examples

3 *M* KCl Aqueous Solution

$$1 \times 3 \times 73.5 \cong 1 \times 3 \times 76.4$$

$$220.5 \cong 229.2$$

0.1 *M* $Et_4N^+NO_3^-$ in Methanol

$$1 \times 0.1 \times 60.4 \cong 1 \times 0.1 \times 60.5$$

$$6.04 \cong 6.05$$

The limiting equivalent conductance data for different cations and anions are listed in Table 3.6 for aqueous, methanol, and ethanol solutions. Note that the limiting equivalent conductance values for the hydrogen ion and hydroxide ion are much greater than for other ions. This makes it difficult to achieve equitransference in a strong acid or base filling solution and is often the reason that slower response is observed in this type of sample. When a liquid junction is placed in a strong acid or base sample, very little diffusion of hydrogen or hydroxide ions into the junction can give rise to unequal transference and sizeable liquid junction potential.

Any electrolyte solution which meets the required conditions can be used in a salt bridge.

One useful filling solution for nonaqueous measurements is methanol saturated with potassium chloride. Since potassium chloride is used as the electrolyte, the filling solution may be placed directly in the reference electrode salt bridge, and an auxiliary salt bridge is not required as in the case of most other filling solution electrolytes.

If the filling solution in contact with the internal element is altered, the standard potential E_9 (Figure 1.3) will change. The stability of this potential may be more sensitive to temperature, and regulation may be required. Also, it takes time to establish the new potential, and the reference electrode should be left overnight before using. Thus, the type of filling solution within a reference electrode is not changed frequently, and the reference electrode

TABLE 3.6

Limiting Equivalent Conductance[a]

Cation	Aqueous	Methanol	Ethanol
Ag^+	61.9	—	17.9
Ba^{2+}	63.6	62.0	—
Ca^{2+}	59.5	61.0	—
Cu^{2+}	53.6	—	—
H^+	349.8	141.8	57.4
K^+	73.5	52.4	22.0
Li^+	38.7	39.8	15.0
Mg^{2+}	53.1	59.0	—
Na^+	50.1	45.9	18.9
NH_4^+	73.5	57.9	19.6
$(CH_3)_4N^+$	44.9	70.1	28.3
$(C_2H_5)(CH_3)_3N^+$	40.8	—	—
$(C_4Hg)(CH_3)_3N^+$	33.6	—	—
$(C_2H_5)_4N^+$	32.7	60.4	27.8
$(C_3H_7)_4N^+$	23.4	46.1	—
$(n\text{-}C_4H_9)_4N^+$	19.5	39.1	—

Anion			
Br^-	78.14	56.4	26.0
Cl^-	76.4	51.2	24.3
CO_3^{2-}	69.3	—	—
ClO_4^-	67.3	70.1	33.5
F^-	55.4	—	—
HCO_3^-	44.5	—	—
I^-	76.8	62.7	28.8
NO_3^-	71.4	60.5	28.0
OH^-	198.6	—	—
SO_4^{2-}	80.0	—	—
SCN^-	66.0	60.8	29.7
Acetate$^-$	40.9	53.0	—
Benzoate$^-$	32.4		
N-Butyrate$^-$	32.6		
Oxalate^{2-}	74.2		
Picrate$^-$	30.4	47.0	26.3
Propionate$^-$	35.8	—	21.0

[a] From R. Parsons, "Handbook of Electrochemical Constants," Butterworth, London, 1959, and L. Meites, "Handbook of Analytical Chemistry," McGraw-Hill, New York, 1973.

with an altered solution is normally an electrode dedicated to a particular application.

The quaternary ammonium salts are often used as electrolytes in nonaqueous solvents since they are much more soluble than the inorganic electrolytes.

Some common solvents in which tetrabutylammonium bromide (Bu_4NBr) and tetraethylammonium perchlorate (Et_4NClO_4) are soluble are:

Bu_4NBr

Pyridine
Dimethylformamide
Acetonitrile
DMSO

Et_4NClO_4

Formamide
Acetone
Acetonitrile
DMSO

In general, more precise selection of an electrolyte as outlined and careful choice of solvent can reduce the junction potential. For example, an auxiliary salt bridge containing 50% dimethylsulfoxide (DMSO), 50% methanol, and a tetraethylammonium perchlorate electrolyte could be used as an intermediate solution between a reference electrode containing methanolic–KCl filling solution and a sample in DMSO solvent. Figure 3.9 illustrates use of an intermediate electrolyte in a salt bridge.

3.2.5 Performance

There are two good methods for testing the reference electrode performance. One test is based on comparing the questionable reference to the other reference known to have good performance. Since one reference electrode potential is offset by the other electrode potential, the test is often referred to as bucking. This

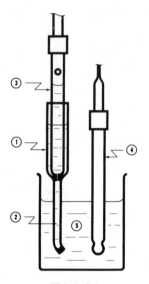

Figure 3.9
Use of Auxiliary Salt Bridges

(1) Salt bridge
(2) Intermediate electrolyte filling solution
(3) Reference electrode
(4) Glass electrode
(5) Sample

test will detect a faulty internal element, a contaminated filling solution, or a very large junction potential.

The other test is specific to the junction and measures its resistance. This test will detect whether the junction is slightly clogged and requires clearing. An ordinary ohmmeter is used to measure the resistance value.

The procedures for these tests are now described.

A. Bucking Test

When reference electrodes are manufactured, they are tested to determine if the standard potential is produced; that is, a calomel internal in saturated potassium chloride solution produces

244 mV and a silver–silver chloride internal produces 199 mV versus the standard hydrogen electrode within a tolerance specification of about ±5 mV at 25°C. This specification may be tested by comparing two reference electrodes; that is, one electrode offsets the other and the difference in potential is observed on the meter. If both reference electrodes have the same internals, zero millivolts within ±5 mV should be observed. If one electrode has a calomel internal and the other has a silver–silver chloride internal, 45 mV within ±5 mV should be observed. These values would be observed on a pH meter in the millivolt mode if both electrodes are functioning properly.

The steps to perform this comparison test are:

(a) Locate zero millivolts on a pH meter which has readability of at least 5 mV. For some meters this may require shorting the inputs using a terminal connector and shorting strap as described in Section 7.1 (Figure 7.1). Then with the readout activated, adjust the standardize control to a point on the scale designated as zero millivolts. On other meters, zero millivolts can be located when in the stand-by mode where the input is shorted and the readout is active. Still other models provide a millivolt absolute mode for convenient pushbutton calibration to zero millivolts.

(b) Connect the reference electrodes to a glass and reference inputs. The shorting strap is removed, but the terminal connector remains and allows the pin jack found on most reference electrode cables to be connected into the glass input. The questionable or comparison reference electrodes may be connected to either input. Only the sign of the observed potential will change when the reference connectors are switched to opposite inputs.

(c) Place both reference electrodes in a beaker containing saturated potassium chloride solution as shown in Figure 3.10. Activate the readout and observe the potential reading. Deactivate the meter, rinse the electrode tips, and place them in a beaker containing pH 7 buffer solution. Repeat these steps using a sample solution.

The potential observed in each solution should be zero millivolts

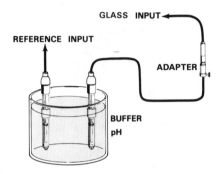

Figure 3.10
Bucking Reference Test

within the tolerance specification which is typically ±5 mV for two reference electrodes of the same internal.

A reference electrode with a calomel internal element may be used to check an electrode with a silver–silver chloride internal element and vice versa. If the silver–silver chloride electrode is connected to the glass electrode terminal and the calomel electrode is connected to the reference terminal, a reading of −45 mV ±5 mV should be observed. For the reverse connections, a reading of +45 mV should be observed. If the questionable reference electrode has a broken internal, normally a large potential shift, greater than 100 mV, would be observed. If a slight potential difference is observed for any of the solutions, greater than the tolerance specification, the junction may have a large liquid junction potential.

B. Junction Resistance Test

A more definitive test for a clogged junction is a resistance measurement. A reference junction which exhibits the typical difficulties encountered with a large liquid junction potential (see Figure 3.8) may have a large resistance. When a reference electrode is made, the manufacturer will often test the resistance of the junction. A new asbestos or quartz fiber junction may have a resistance of 5000 ohms. A new annular ceramic or sleeve junction has less resistance. With use and aging, the resistance value often increases. A junction which exhibits higher than 20,000 ohms should undergo the procedure for clearing a junction. This is only an approximate resistance value for which a user may wish to initiate the clearing procedure. Most clogged junctions exhibit much higher resistance values.

The resistance measurement is made using a standard ohmmeter. It consists of making contact with the reference internal filling solution and an external solution in which the junction is immersed. The steps are:

(a) Prepare the ohmmeter. Place the range switch on 1K. Make one lead flexible by attaching a wire to the probe point. Short the leads together and adjust zero on the ohmmeter scale (see Figure 3.11).

(b) Pour sufficient reference electrode filling solution into a beaker to allow immersion of the reference junction. Place the two probes from the ohmmeter into the solution to see the approximate solution resistance.

(c) Place the flexible lead from the ohmmeter into the reference electrode filling hole and make contact with the internal filling solution.

(d) Place the reference electrode tip and the other probe from the ohmmeter into the beaker of filling solution (see Figure 3.11).

(e) Subtract the solution resistance as determined in Step (b) from this reading to determine the junction resistance.

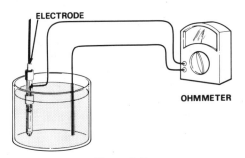

Figure 3.11
Reference Junction Resistance Test

This test is a test of the junction only. It does not test the internal, cable lead, or filling solution purity. It does, however, focus on the junction condition, which is typically a source of difficulty. Also, it does not require a second reference electrode, which is assumed to be functioning properly.

C. Unclogging a Liquid Junction

If a reference junction has a high resistance value or is exhibiting the typical problem of liquid junction potentials, such as readings which are drifting or require a long stabilization time, it may require clearing. There are a number of steps which can be implemented to lower the junction resistance and liquid junction potential:

(1) The easiest step is to replace the filling solution and allow the junction to soak overnight in 0.1 M KCl solution. This step is good for a junction which has not been recently used or has been allowed to dry out. The filling solution inside the reference electrode is dispensed by shaking it out through the filling hole. If crystals of potassium chloride are impacted at the electrode tip, add distilled or deionized water and allow the crystals to dissolve. Several rinsings with water may be necessary to dissolve the crystals quickly. The first addition of deionized water to a silver–silver chloride reference salt bridge will cause a white precipitate to form as complexes of silver chloride are precipitated. This does not harm the electrode and is eliminated by further rinsings. After the encrusted KCl is removed by dissolution, add fresh filling solution and place the electrode tip in 0.1 M KCl for overnight soaking.

(2) The second step is to apply a slight pressure to the filling hole or vacuum to the junction tip in order to force filling solution through the junction. This can be accomplished using an air or nitrogen gas line or an aspirator hose (see Figure 3.12). Only a partial seal of the hose to the electrode is needed since about an atmosphere of pressure is being applied.

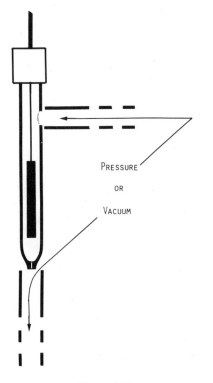

Figure 3.12
Unclogging the Junction

When neither of these steps produces a clear junction as evidence by the resistance test, there are some other more involved steps which can be taken; these steps are considered more severe, since they can limit the electrode life.

(3) One such step involves boiling the junction for 10 minutes in dilute KCl solution. First dilute a saturated potassium chloride solution about 1:10 with water. Place the reference electrode tip in the boiling solution and bring to a boil for about 10 minutes. Do not allow the electrode to remain in the boiling solution too long (evidenced by a fogging of

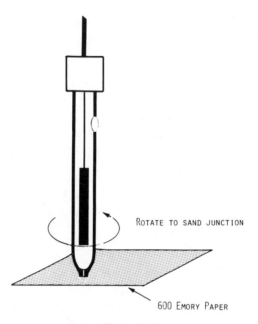

Figure 3.13
Reducing Junction Resistance

glass). Allow the electrode to cool while immersed in the
solution before testing again.
(4) If Step (3) does not clear the junction, as a last step, it may
be sanded with #600 emory paper. This is a last resort,
since sanding limits the electrode life. A drop of water is
placed on the emory paper which is lying on a flat surface.
The junction is then rotated perpendicular to the paper in a
circular fashion (see Figure 3.13). This procedure is continued with periodic resistance tests until the desired ohm
value is obtained.

3.3 COMBINATION ELECTRODES

As the name implies, a combination electrode is a combination of
the glass and reference electrodes into a single probe. In fact, a

combination electrode may be used as either a reference or glass electrode simply by connecting only the appropriate connector to the pH meter. If the glass electrode only is used, the reference electrode connector should be connected to solution ground to shield the electrode. The combination electrode is constructed with the reference surrounding the glass as shown in Figure 3.14.

The functioning and parameters described in the previous sections of this chapter for glass and reference electrodes apply to the combination electrode. The combination is normally constructed with a silver–silver chloride internal reference portion. With this internal the electrode diameter can be small; with a calomel internal a larger diameter would be required.

The reference junction material may vary. An annular ceramic junction which surrounds the glass bulb is often preferred since it provides numerous filling solution leak paths and a stable potential in most samples.

The combination electrode shielding normally extends over only

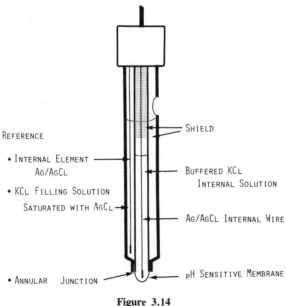

Figure 3.14
Combination Electrodes

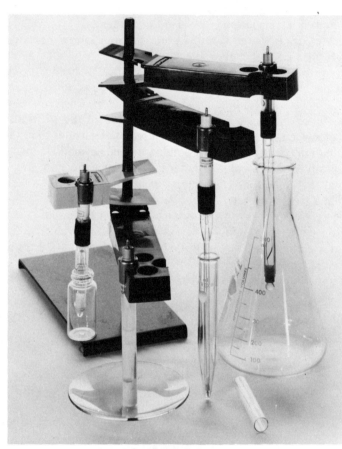

Figure 3.15
Small-Volume Sample Containers

a portion of the electrode body and relies on the reference filling solution to complete the shielding. Therefore, as previously mentioned, the filling solution should be at a level which completes the shielding.

The main advantage in using a combination electrode is with the measurement of small volume samples or samples in limited access containers as shown in Figure 3.15.

3.4 ELECTRODE SELECTION

There are some basic decisions to be made when selecting electrodes for an application. These include

(1) combination or electrode pair,
(2) type of pH glass,
(3) type of reference junction,
(4) type of reference internal element.

The first decision becomes a comparison of the advantages. The combination electrode provides simplicity and capability of measuring small-volume samples. Also, the close proximity of the reference junction to the glass bulb is an advantage in high resistance solutions, such as distilled water or nonaqueous solutions, since a more stable reading may result.

On the other hand, an electrode pair has the advantages of flexibility in reference electrode choice and reduced replacement cost. Different junction types, including a remote salt bridge, may be used with the same glass electrode in order to determine which reference provides the best response. Normally, the selection of junction types of a combination electrode is limited. There is also more flexibility of internal elements since most combinations are made with silver–silver chloride internals. An application which must exclude silver ion contamination may require an electrode pair. If one of the electrodes of a pair is broken, its replacement cost is less than replacing a combination electrode.

Thus, the advantages and limitations when applied to the application should dictate the selection of electrode type.

The next decision is a choice of pH glass designed for use over the 0–11 pH or the 0–14 pH range. The former is preferred because of its lower resistance which provides a more stable reading and faster response. If the application involved pH measurements above pH 11, however, a full-range glass should be employed.

The last two decisions are described in Section 3.2. Table 3.3 provides a general indication of the preferred junction for various types of samples. The advantages and limitations of the silver–silver chloride and the calomel internals are also described.

3.5 STORAGE OF ELECTRODES

Since the pH measurement is dependent on a hydrated glass bulb and a free-flowing junction from the reference electrode, keeping both electrodes wet is vital.

Fast response by the glass electrode to sample measurements has been observed when the electrode is first soaked in a slightly acid solution. Some users soak the glass bulb in distilled water. This low ionic strength solution extracts the ions from the bulb, causing a much slower response. Others use pH 7 phosphate buffer which, over a long time period, ages the electrode slightly. A pH 4 buffer soaking and storage solution, however, provides a fast response for most sample pH measurements.

In order to provide a fast-responding and low junction potential reference electrode, the junction must remain unclogged with low resistance and with flow of filling solution. If the junction is allowed to dry, some sample particles or precipitate may be difficult to remove from the junction by soaking.

One of the best solutions for soaking a reference junction would be its own electrolyte filling solution. Its tendency to creep and leave a salt crust, however, makes it a messy storage solution. By diluting ($\sim$1:100) this saturated potassium chloride solution, its messy nature is reduced. By providing a potassium chloride solution on both sides of the junction during storage, a freer flowing junction is obtained for the next measurement.

The internal filling solution should be at a height above the storage solution (or sample solution when measuring) in order to provide a head pressure and positive flow of filling solution into the storage solution or sample. The sleeve or plug should cover the filling hole to reduce the flow of electrolyte. This sleeve, plug or tape, however, must be removed from the filling hole during a measurement.

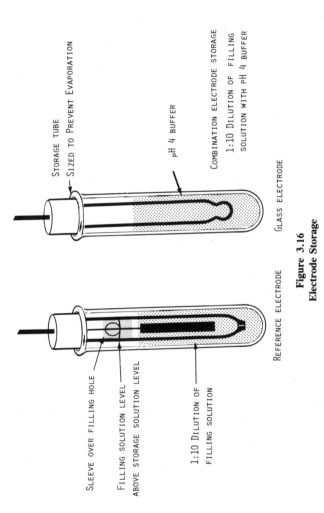

Figure 3.16
Electrode Storage

The combination electrode can be stored in a combination of the recommended glass and reference storage solutions; that is, pH 4 buffer with a few drops of saturated potassium chloride added.

A convenient means of storing electrodes is in test tubes. The test tube size should be selected so as to have a large enough diameter to accommodate the electrode but not the cap. This allows the electrode to be supported by its cap and prevents the glass bulb from contacting the bottom of the test tube and also minimize evaporation of the storage solution (see Figure 3.16).

Chapter 4

Standard Solutions (Buffers)

Since a glass or reference electrode may vary in the potential that it produces in a given solution, standard solutions are required in order to equate different electrode pairs at a given pH. These standard solutions are referred to as buffers. As the name implies, these solutions have the ability to resist change in pH due to dilution or contamination.

The pH of an unknown solution is determined by comparison to an accepted standard solution. In other words, the pH(X) for an unknown solution can be determined by relating the potential E_x obtained with a pair of electrodes in the unknown solution to the potential E_s obtained in an accepted standard of pH(S) if the electrode response S (slope) is known. The pH meter displays this difference observing the expression

$$\text{pH}(X) = \text{pH}(S) - \frac{E_x - E_s}{S}$$

4.1 CHARACTERISTICS

There are many different characteristics which determine if a particular solution composition will qualify as a buffer. The primary parameters are the buffering capacity, the effect dilution has on the buffer, and the pH change with temperature. Other characteristics which must be considered are shelf life, growth of mold, and absorption of carbon dioxide.

4.1.1 Buffering Capacity, β

The buffering capacity, sometimes referred to as buffer value, is a measure of the ability of the solution to resist pH change when a

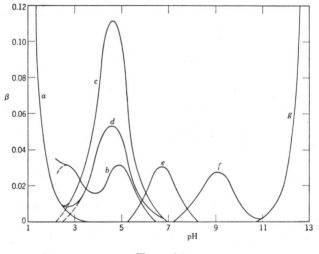

Figure 4.1
Buffering Capacity

Buffer values β for solutions of strong acid and alkali and for buffer solutions composed of weak acids and strong alkali, as a function of pH:

(a)	HCl	(e)	$0.05\ M$ KH_2PO_4 + NaOH
(b)	$0.05\ M$ phthalic acid + NaOH	(f)	$0.05\ M$ H_3BO_3 + NaOH
(c)	$0.2\ M$ acetic acid + NaOH	(g)	NaOH
(d)	$0.1\ M$ acetic acid + NaOH		

strong acid or base is added to the buffer solution. This value is unity when 1 liter of buffer solution will take up 1 gram-equivalent of strong alkali or strong acid per unit increase in pH, assuming no volume change. The buffering ability of a weak acidic or basic group is limited to pH = $pK_a \pm 1$ with the greatest capacity at pH = pK_a. For example, a liter of pH 4.01 buffer solution will require up to 0.016 gram-equivalent of strong acid before the pH value is decreased to pH 3.01. The higher the buffer value β of the solution, the greater its capacity to buffer. (See Figure 4.1.)

4.1.2 Dilution Value, $pH_{\frac{1}{2}}$

The dilution value is the change in pH of a buffer solution when diluted with an equal quantity of water. This value is positive when the pH increases with dilution and negative when the pH de-

TABLE 4.1

Buffers

Solution composition (molality)	pH at 25°C	Temperature range (°C)	Dilution value $\Delta pH_{\frac{1}{2}}$	Buffer value β	Temperatre coefficient dpH/dt
Primary					
KH tartrate (saturated at 25°C)	3.557	25–95	—	—	—
0.05 m Citrate (KH$_2$C$_6$H$_5$O$_7$)	3.776	0–50	+0.024	0.034	−0.0022
0.05 m Phthalate (KHC$_8$H$_4$O$_4$)	4.008	0–95	+0.052	0.016	+0.0012
0.025 m KH$_2$PO$_4$ + 0.025 m Na$_2$HPO$_4$	6.865	0–95	+0.080	0.029	−0.0028
0.008695 m KH$_2$PO$_4$ + 0.03043 m Na$_2$HPO$_4$	7.413	0–50	+0.07	0.16	−0.0028
0.01 m Na$_2$B$_4$O$_7$·10H$_2$O	9.180	0–95	+0.01	0.020	−0.0082
0.025 m NaHCO$_3$ + 0.025 m Na$_2$CO$_3$	10.012	0–50	+0.079	0.029	−0.0096
Secondary					
0.05 m KH$_3$(C$_2$O$_4$)$_2$·H$_2$O	1.679	0–95	—	—	—
Ca(OH)$_2$ (saturated at 25°C)	12.454	0–60	−0.28	0.09	−0.033

creases. The smaller the $\Delta pH_{\frac{1}{2}}$ value, the less sensitive the buffer is to change in the concentration of its components. Examples of this value are +0.052 pH increase in a 0.05 m concentration of pH 4.01 buffer, while an increase of +0.01 pH is observed with a 0.01 m concentration of pH 9.18 buffer when diluted by an equal volume of water. (See Table 4.1.)

4.1.3   Temperature Effect, dpH/dT

A change in temperature changes the buffer value by affecting the activity of the hydrogen ion. This temperature coefficient expresses the pH change per degree centigrade. A calcium hydroxide buffer has a large coefficient of approximately −0.033 pH unit for each degree centigrade higher temperature. (See Tables 4.1 and A.6.)

4.2   BUFFER COMPOSITION—PRIMARY

Although there are numerous types and compositions of buffers, three common primary buffers are most frequently used. The word *primary* implies "to be preferred over other buffers of similar pH value and affording greater accuracy." These buffers are the pH 4.01 phthalate buffer, the pH 6.86 phosphate buffer, and the pH 9.18 borax buffer.

The pH 4.01 buffer has a composition of 0.05 m potassium hydrogen phthalate ($KHC_8H_4O_4$). It has a relatively low buffering capacity but is stable to dilution and change in temperature. It should be prepared with CO_2-free distilled water. It is subject to mold growth but is normally inhibited by fungicide.

The pH 6.86 buffer has a composition of 0.025 m potassium dihydrogen phosphate (KH_2PO_4) and 0.025 m disodium hydrogen phosphate (Na_2HPO_4). It is fairly stable to contamination and moderately affected by temperature and dilution.

The pH 9.18 buffer has a composition of 0.01 m sodium tetraborate decahydrate ($Na_2B_4O_7 \cdot 10H_2O$). In solution this salt hydrolyzes to form boric acid and sodium borate. It has good stability toward dilution. It should be protected from atmospheric CO_2 for maximum accuracy; however, it is surprisingly stable in an open beaker.

Table 4.1 lists the primary buffers and two secondary buffers that are recommended by the National Bureau of Standards. The citrate buffer was developed to replace the tartrate buffer with better stability and for use when the phthalate buffer could not be used because of its absorption spectrum. The pH 10 carbonate buffer is being used with increasing frequency to replace the borax buffer. Although it has a higher sensitivity to dilution, it does not have a tendency to polymerize as does borax, and consequently the ionic strength remains stable. The pH 7.413 buffer was developed for physiological pH measurements and is rapidly being replaced by tris buffer (discussed in Section 4.4.5).

Because of the well-known abnormalities of the liquid junction potentials when making pH measurements at the extremes of the pH scale, the tetroxalate and calcium hydroxide buffers are regarded as secondary. Both are sensitive to dilution, but only pH 12.45 buffer is highly sensitive to temperature changes.

4.3   OTHER BUFFERS

Many other buffers have been developed for various purposes. Generally, these alternate compositions are less stable or accurate than the primary buffers described the the National Bureau of Standards. They are often employed when special conditions must be met or when the buffer is more compatible with the sample. Some of these buffers are described in Section 4.4. One scheme of buffer compositions which cover the pH range 1–10 has been developed by Clark and Lubs. These are described in Table A.3 and, generally, with others listed in the "Handbook of Biochemistry" [5].

References

1. Bates, R. G., Revised Standard Values for pH Measurements from 0 to 95°C, *J. Res. Nat. Bur. Std.* **66A,** 179 (1962).

2. Bates, R. G., Standardization of Acidity Measurements, *Anal. Chem.* **40**, 28A (1968).
3. Bower, V. E., and Bates, R. G., Standards for pH Measurements from 60° to 95°C, *J. Res. Nat. Bur. Std.* **59**, 261 (1957).
4. Bower, V. E., and Bates, R. G., pH Value of the Clark and Lubs Buffer Solutions at 25°C, *J. Res. Nat. Bur. Std.* **55**, 197 (1955).
5. H. A. Sober, ed., "Handbook of Biochemistry," 2nd ed., Molecular Biology, p. J195. Chem. Rubber Co., Cleveland, Ohio, 1970.
6. Perrin, D. D., and Dempsey, B., "Buffers for pH and Metal Ion Control." Halsted Press, New York, 1974.

4.4 SPECIAL BUFFERS

There are a number of buffer standards which have become common for a specific pH measurement application. These include photographic developer, heavy water (D_2O), nonaqueous, seawater, and biological pH measurements.

4.4.1 Photographic

This buffer is a hybrid calcium hydroxide with a pH value of 11.88 at 25°C. It has most of the characteristics of the pH 12.45 buffer such as high sensitivity to temperature changes, but it is at a pH value exactly where several popular developers should be maintained. This buffer has high ionic strength because it contains 2 M sodium chloride.

4.4.2 Heavy Water (D_2O)

Standards have been developed by the National Bureau of Standards to cover the 4.3 to 10.7 pD range for use in nuclear technology and research in biomedical fields.

Ordinary aqueous standards can be used to measure the pD of samples containing deuterium activity. However, a constant correction must be applied, and considerable response time is required to replace the H_2O with D_2O completely in the outer gel layer of

TABLE 4.2

D_2O Standards; Standard Reference Values pD(S) for the Measurement of Acidity (pD) in Heavy Water

t (°C)	Solution (molality)		
	0.05 KD_2 citrate	0.025 KD_2PO_4 + 0.025 Na_2DPO_4	0.025 $NaDCO_3$ + 0.025 Na_2CO_3
5	4.378	7.537	10.996
10	4.352	7.504	10.924
15	4.330	7.475	10.856
20	4.310	7.450	10.794
25	4.293	7.429	10.737
30	4.279	7.411	10.684
35	4.268	7.397	10.637
40	4.259	7.386	10.595
45	4.253	7.380	10.558
50	4.250	7.377	10.527

the glass. The correction between a pD value and standardization with aqueous buffers is shown by the equation

$$pD = pH_0 + 0.40$$

where pH_0 is the observed pH value.

When pD buffers are used, an uncertainty of less than 0.02 pD can be obtained. Three standard reference pD buffer values versus temperature are listed in Table 4.2. They are basically the citrate, phosphate, and carbonate compositions used as aqueous pH buffers but with deuterium replacing the hydrogen.

References

1. Baumann, E. W., Determination of Deuterium Ion Concentration in Dilute Unbuffered Deuterium Oxide Solutions, *Anal. Chem.* **38,** 1255 (1966).
2. Covington, A. K., Paabo, M., Robinson, R. A., and Bates, R. G., Use of the Glass Electrode in Deuterium Oxide and the Relation between the Standardized pD (pa_D) Scale and the Operational pH in Heavy Water, *Anal. Chem.* **40,** 700 (1968).
3. Forcé, R. K., and Carr, J. D., Temperature-Dependent Response of the Glass Electrode in Deuterium Oxide, *Anal. Chem.* **46,** 2049 (1974).

4. Gary, R., Bates, R. G., and Robinson, R. A., *J. Phys. Chem.* **68**, 1186, 3806 (1964); **69**, 2750 (1965).

4.4.3 Nonaqueous

The glass electrode responds to hydrogen ion activity when immersed in a mixture of solvents of which water is at least a few percent. With nonaqueous measurements, however, significant error can occur due to the medium effect, discussed in Chapter 1, on the activity and on the liquid junction of the reference electrode giving rise to large liquid junction potentials.

Measurement scales based on the activity of a hydrogen ion are limited to a specific solvent or solvent mixture. Therefore, comparison of hydrogen ion activity in an aqueous buffer with that in a nonaqueous solvent mixture does not have quantitative significance. In order to provide an operational scale of pH in nonaqueous solvents, the National Bureau of Standards has developed buffers by relating the mass law to the pK value of the weak acid or base which controls the acidity of different solvents.

Three nonaqueous buffers defined by the National Bureau of Standards, which contain 50% (by weight) methanol, result in higher pH values than aqueous solutions of the same solutes and molalities. These buffers are defined in Table 4.3.

A normal calomel reference electrode may be used in solution containing up to 68% methanol with a large but stable junction potential. A means of reducing this potential is to change the filling solution as discussed in Chapter 3. For example, a methanol solvent saturated with potassium chloride may provide a more stable reference electrode. This error is also discussed in Section 6.2.1.E.

References

1. Bates, R. G., Paabo, M., and Robinson, R. A., Interpretation of pH Measurements in Alcohol Water Solvents, *J. Phys. Chem.* **67**, 1833 (1963).
2. Paabo, M., Robinson, R. A., and Bates, R. G., Reference Buffer Solutions for pH Measurements in 50% Methanol. Dissociation Constants of Acetic Acid and Dihydrogen Phosphate Ion from 10 to 40°C, *J. Am. Chem. Soc.* **87**, 415 (1965).

TABLE 4.3

Nonaqueous Standard; Standard Reference Values pH*(S) for the Measurement of Acidity (pH*) in 50 wt % Methanol–Water

t (°C)	0.02 HAc,[a] 0.02 NaAc, 0.02 NaCl	0.02 NaHSuc,[b] 0.02 NaCl	0.02 KH₂PO₄, 0.02 Na₂HPO₄, 0.02 NaCl
	Solution (molality)		
10	5.560	5.806	7.937
15	5.549	5.786	7.916
20	5.543	5.770	7.898
25	5.540	5.757	7.884
30	5.540	5.748	7.872
35	5.543	5.743	7.863
40	5.550	5.741	7.858

[a] Ac = acetate.
[b] Suc = succinate.

3. Woodhead, M., Paabo, M., Robinson, R. A., and Bates, R. G., Acid–base Behavior in 50-Percent Aqueous Methanol, *J. Res. Nat. Bur. Std.* **69A,** 263 (1965).

4.4.4   Seawater

Seawater is a constant high ionic medium. If the glass electrode is transferred from dilute buffers of less than 0.1 m ionic strength to seawater normally have an ionic strength exceeding 0.6 m ionic strength, problems with the measurement are observed. The difference in mobilities of the ions in seawater plus the high concentrations of these ions cause a large liquid junction potential at the reference electrode. Also, the difference in the ionic activity between the seawater sample and the glass electrode internal solution cause a slight changing asymmetrical potential. Thus, the result is a slow responding measurement which is in error.

In order to minimize these problems, a secondary standard consisting of artificial seawater buffered with tris may be used. The effect that increased salinity has on the activity of the hydrogen ion

TABLE 4.4

Seawater Standard

	Molality	Temperature (°C)	pH
NaCl	0.4186	5	8.835
MgCl$_2$	0.0596	15	8.517
Na$_2$SO$_4$	0.02856	25	8.224
CaCl$_2$	0.005	35	7.953
Ionic strength	0.66		
Plus buffering agent	0.02 m tris-(hydroxymethyl)aminomethane 0.02 m tris-HCl		

in the buffer is offset by the residual liquid junction potential. Table 4.4 lists the composition of artifical seawater buffers and the pH values at various temperatures.

References

1. Bates, R. G., and Macaskill, J. B., Analytical Methods in Oceanography, *Adv. Chem. Ser.* No. 147, 110 (1975).
2. Hansson, I., *Deep-Sea Res.* **20**, 489 (1973).
3. Hawley, J. E., and Pytkowicz, R. M., *Mar. Chem.* **1**, 245 (1973).
4. Kester, D. R., and Pytkowicz, R. M., *Limnol. Oceanogr.* **12**, 243 (1967).
5. Khoo, K. H., Ramette, R. W., Culberson, C. H., and Bates, R. G., Determination of Hydrogen Ion Concentrations in Seawater from 5 to 40°C: Standard Potentials at Salineties from 20 to 45%, *Anal. Chem.* **49**, 20 (1977).
6. Mehrbach, C., Culberson, C. H., Hawley, J. E., and Pytkowicz, R. M., *Limnol. Oceanogr.* **18**, 897 (1973).
7. Pytkowicz, R. M., Ingle, S. E., and Mehrbach, C., *Limnol. Oceanogr.* **19**, 665 (1974).
8. Pytkowicz, R. M., and Kester, D. R., *Oceanogr. Mar. Biol. Ann. Rev.* **9**, 11 (1971).
9. Pytkowicz, R. M., Kester, D. R., and Burgener, B. C., *Limnol. Oceanogr.* **11**, 417 (1966).
10. Smith, W. H., Jr., and Hood, D. W., "Recent Researches in the Fields of Hydrosphere, Atmosphere, and Nuclear Geochemistry," p. 185. Maruzen, Tokyo, 1964.
11. Strickland, J. D. H., and Parsons, T. R., "A Practical Handbook of Seawater Analysis." Fisheries Res. Board of Canada, Ottawa, 1968.

4.4.5   Biological Buffers

Electrodes used to make pH measurements in biological media are often standardized with tris buffer. One composition of this buffer is

0.0500 mol/kg tris-(hydroxymethyl)aminomethane hydrochloride

0.01667 mol/kg tris-(hydroxymethyl)aminomethane

At 37°C this buffer has a pH value of 7.382 with a temperature coefficient of -0.026 pH. It is used with physiological measurements because of its compatibility and solubility with this type of sample. It does not appear to inhibit many enzyme systems and has a temperature coefficient near that of blood (-0.015 pH/°C).

The disadvantages of tris buffer are that it is a primary aliphatic amine which can react with some samples, that it does not have a large buffering capacity at this pH value, and that it appears to be reactive with linen fiber reference junctions. Most often, these disadvantages are not significant or can be overcome by use of another type of reference junction.

In addition to its compatibility with biological samples, tris buffer has several other advantages over standard inorganic buffer salts. It does not precipitate calcium salts or absorb carbon dioxide appreciably. It is nonhygroscopic and has good stability.

Accurate, reliable pH measurements can be made in a tris-chloride buffer and solutions by following the recommendations and procedures listed here:

(1)   Use an electrode pair or a combination electrode with a ceramic, quartz, or sleeve junction.

(2)   Prepare tris buffer with known purity tris and high purity, standardized HCl. Make up with nitrogen-purged deionized or distilled water.

(3)   Allow the pH reading to stabilize. This may require as long as 5 minutes for the first measurement or an occasional measurement in high concentration buffers. Stabilization time is reduced as successive measurements are made.

TABLE 4.5

BES, Tricine, and Tris Buffers

Temperature (°C)	1:1 BES + NaBESate		0.06 m Tricine and 0.02 m sodium tricinate	0.0167 m tris + 0.05 m tris-HCl
	0.02 m	0.1 m		
5	7.448	7.459	8.023	8.303
10	7.361	7.370	7.916	8.142
15	7.284	7.290	7.813	7.988
20	7.204	7.213	7.713	7.840
25	7.134	7.134	7.621	7.698
30	7.063	7.064	7.527	7.563
37	6.969	6.966	7.407	7.382
40	6.926	6.925	7.355	7.307
45	6.866	6.858	7.275	7.186
50	6.800	6.794	7.197	7.070

(4) In order to obtain the most accurate measurements, all solutions should be temperature controlled. The best results will be obtained if the sample and the buffer are at the same temperature.

(5) Span checks should be made with two tris/tris-chloride buffers to verify condition of the pH (glass) electrode.

The pH values versus temperature of two other frequently used physiological buffers,

BES: *N,N*-bis(2-hydroxyethyl)-2-aminoethanesulfonic acid
         (pH range 6.6–7.4)

tricine: tris-(hydroxymethyl)methylglycine
         (pH range 7.2–8.5)

along with tris are listed in Table 4.5.

The physical properties of some other biological buffers are listed in Table 4.6. These are further described in reference [2].

References

1. Gomori, G., Buffers in the Range of pH 6.5 to 9.6, *Proc. Soc. Exp. Biol. Med.* **62,** 33 (1946).

TABLE 4.6

Physical Properties of Biological Buffers

Structure	Proposed name	pK_a at 20°	pK_a/°C	Saturated solution at 0° (M)	Metal–buffer binding constants log K_M			
					Mg^{2+}	Ca^{2+}	Mn^{2+}	Cu^{2+}
$NHCH_2CH_2SO_3^-$ (morpholine)	MES	6.15	−0.011	0.65	0.8	0.7	0.7	Negl
$NaO_3SCH_2CH_2N$⁺...$NCH_2CH_2SO_3^-$	PIPES	6.8	−0.0085	—	Negl	Negl	Negl	Negl
$H_2NCOCH_2NH_2CH_2CH_2SO_3^-$	ACES	6.9	−0.020	0.22	0.4	0.4	Negl	4.6
$(HOCH_2CH_2)_2=NHCH_2CH_2SO_3^-$	BES	7.15	−0.016	3.2	Negl	Negl	Negl	3.5
$HOCH_2CH_2N$⁺...$NCH_2CH_2SO_3^-$	HEPES	7.55	−0.014	2.25	Negl	Negl	Negl	Negl
$(HOCH_2)_3=CNH_2CH_2COO^-$	Tricine	8.15	−0.021	0.8	1.2	2.4	2.7	7.3
$(HOCH_2)_3=CNH_2$	Tris	8.3	−0.031	2.4	Negl	Negl	Negl	—
$H_3NCH_2CONHCH_2COO^-$	Glyclyglycine	8.4	−0.028	1.1	0.8	0.8	1.7	5.8

2. Good, N. E., Winget, G. D., Winter, W., Connolly, T. N., Izawa, S., and Singh, R. M. M., Hydrogen Ion Buffers for Biological Research, *Biochemistry* **5**, 467 (1966).
3. Woodhead, M., Paabo, M., Robinson, R. A., and Bates, R. G., Buffer Solutions of Tris-(Hydroxymethyl) Aminomethane for pH control in 50 Weight Percent Methanol from 10° to 40°C, *Anal. Chem.* **37**, 1291 (1965).
4. Durst, R. A., and Staples, B. R., Tris/Tris-HCl: A Standard Buffer for Use in the Physiologic pH Range, *Clin. Chem.* **18**, 206 (1972).
5. Bates, R. G., Roy, R. N., and Robinson, R. A., Buffer Standards of Tris-(Hydroxymethyl) Methylglycine ("Tricine") for the Physiological Range pH 7.2 to 8.5, *Anal. Chem.* **45**, 1663 (1973).
6. Fossum, J. H., Markunas, P. C., and Riddick, J. A., Tris-(Hydroxymethyl) Aminomethane as an Acidimetric Standard, *Anal. Chem.* **23**, 491 (1951).
7. Westcott, C. C., and Johns, T., pH Measurements in Tris Buffer, Beckman Instruments Tech. Rep. 542 (1971).
8. Woodhead, M., Paabo, M., Robinson, R. A., and Bates, R. G., Acid–Base Behavior in 50-Percent Aqueous-Methanol Thermodynamics of the Dissociation of Protonated Tris-(Hydroxymethyl) Aminomethane and Nature of the Solvent Effort, *J. Res. Nat. Bur. Std.* **69A**, 263 (1965).
9. Roy, R. N., Swensson, E. E., LaCross, G., and Krueger, C. W., Standard Buffer of *N,N*-Bis(2-Hydroxyethyl)-2-Aminoethanesulfonic Acid (BES) for Use in the Physiological pH Range 6.6 to 7.4, *Anal. Chem.* **47**, 1407 (1975).
10. Koch, W. F., Biggs, O. L., and Diehl, H., Tris-(Hydroxymethyl) Aminomethane a Primary Standard, *Talanta* **22**, 637 (1975).
11. Bower, V. E., Paabo, M., and Bates, R. G., pH Standard for Blood and Other Physiologic Media, *Clin. Chem.* **7**, 292 (1961).
12. Hill, D. E., and Spivey, H. O., Errors in pH Measurement in Dilute or Sulfhydryl-Containing Buffers, *Anal. Biochem.* **27**, 500 (1974).
13. Stormorken, H., and Newcomb, T. F., The Use of Buffers in Blood Coagulation Studies, *Scand. J. Clin. Lab. Invest.* **8**, 237 (1956).

4.5   VERIFICATION OF A BUFFER

Most buffers prepared with CO_2-free distilled water and with the salts accurately weighed will be precise for a considerable time. Over a period of time, however, many buffers are subject to mold growth, they polymerize, or are affected by pick-up of carbon dioxide. In the proper container, normally glass or polyethylene, sealed from contamination, a standard NBS-type buffer should last at least 6 months. Most manufacturers of buffer solutions include a fungicide, such as butaben, in the buffer to inhibit mold growth.

The only practical method of verifying the accuracy of a buffer is to compare it to another known buffer solution. NBS-certified salts may be used and the NBS procedure followed to produce a primary buffer whose pH value is closest to the pH value of the buffer in question. The comparison of buffers should be made at the same temperature, using the same electrodes and employing good technique.

Chapter 5

pH Measurement Technique

There are many different parameters which can affect the pH measurement, but only two are, normally, sources of significant error. These two parameters are the electrodes and the technique employed. This chapter discusses the components of proper technique and summarizes the various factors of accuracy. The discussion is directed at such questions as what response time can be expected, should the sample be stirred or static, and how does temperature effect the pH reading?

5.1 RINSING

The objective of rinsing the electrodes between measurements is to prevent contamination by carry-over on the electrodes. It has become common practice to use distilled water from a squirt bottle to rinse the electrodes and then to blot them with a tissue. This practice, however, has led to the improper technique of wiping rather than blotting the electrodes. Wiping often results in slower responding electrodes, thus requiring considerable time before a stable pH reading is obtained.

This phenomenon is caused by transfer of static body charge onto the high-resistance glass bulb. In essence, the bulb acts like a charged capacitor, and the charge takes time to dissipate. Large static body charges are generated, for example, on dry, low humidity days when wearing a nylon shirt or lab coat, or standing on a rubber mat. Then, as the operator wipes the glass electrode, the tissue becomes wet, providing a low resistance path for the charge transfer to the bulb. Once a charge is transferred to the bulb, it requires time to dissipate.

The point of this discussion is to blot and not to wipe the electrodes. In other words, do not provide a pathway for charge transfer. Since the rinse is done with distilled water which has very little buffering capacity, any excess water on the electrodes is not going to have a significant effect on the next solution.

The preferred method of rinsing is with a portion of the next sample. This portion is then discarded, and the measurement is taken on another sample portion. Thus, no electrode blotting is necessary since the carry-over is the solution to be measured next. This technique, however, is not always practical because of limited sample volume or the necessity for better electrode cleaning.

If very viscous or sticky samples are being measured, it may take more than a distilled water rinse to remove the sample from the electrode tips. In some cases, a solvent mixture or a series of solvents may be used to rinse the electrodes. For example, the first solvent dissolves the sample, and this rinse is followed by a distilled water rinse to remove the solvent. Remember, the glass bulb must have an aqueous hydrated layer in order to function properly.

5.2 STIRRING

The choice of whether to make the pH measurement with stirred or static conditions should be based on many different parameters, such as sample type, sample volume, and the length of time for which the measurement is being made. In order to provide greater consistency of the reference junction potential, standardization and sample measurements should be done under the same condition; that is, both stirred or both static.

The main advantages to stirring are a more representative pH measurement of a heterogeneous solution, fast response, and the prevention of the slight alkaline effect from the glass bulb dissolution. Often, a viscous, precipitate, or suspension-type sample or a sample which tends to separate into a supernate and sediment is encountered. By stirring this type of sample, the pH value appears homogeneous to the electrodes. By stirring the sample, the solution which was previously around the bulb is more rapidly replaced with the sample, thus providing a faster response. Also, the liquid junction potential equilibrium is more rapidly established in a stirred solution, thus providing a faster response.

If the glass electrode is left in a weakly buffered sample for a long time period, dissolution of the alkaline compounds on the bulb surface causes a slight upward drift of the pH value. Since this upward drift is very slight, it becomes important only when high accuracy is required. This effect can be eliminated by stirring the solution, since the solution surrounding the bulb is constantly being replaced.

The main disadvantages to stirring during a pH measurement are limits on sample volume, possible sample temperature changes, and possible increase in measurement noise. In order to stir a solution, some sort of device (mechanical or magnetic) is needed. This requires space and thus a greater sample volume than would be required if the sample were not stirred. Both the glass and reference must be clear of the magnet or stirrer. If the magnetic stirrer is turned too high, it can lose control of the magnet causing it to jump and result in damage to an electrode.

If the sample beaker is sitting on top of a magnetic stirrer, the heat from the stirrer motor may elevate the sample temperature. This would cause a slow drift of the pH value and may not be identified as a temperature change by the operator. A foam pad between the stirrer platform and the beaker will further isolate the sample and minimize this effect.

A slightly clogged or high resistance junction will have a larger junction potential. If the flow past this junction causes changes in the potential, greater noise in the pH display or recording will be

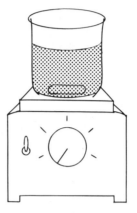

Figure 5.1
Stirring versus Static

Disadvantages of stirring
    Magnet limits sample size and can cause electrode breakage
    Sample may be heated by stirrer motor unless insulated with foam pad
    Noisy reading with flow past high potential junction

Advantages of stirring
    Fast response
    Representative pH value

observed than occurs in the static measurement. Therefore, it is important to have a reference electrode with a low junction potential if the sample is to be stirred. Also, some magnetic stirrers can introduce eddy currents which affect the measured potential. These advantages and disadvantages are summarized in Figure 5.1.

5.3 RESPONSE

When an electrode is functioning properly and has been preconditioned (soaked) in an appropriate storage solution, it should respond to changes in NBS-type buffers (e.g., pH 6.86–4) within 10 seconds to 98% of the final reading. This, of course, is an ideal situation, and in practical applications, the time response depends on the type of sample and the electrodes being used. Low ionic strength, unbuffered, nonaqueous, high salt, and other difficult sample solutions may require much more time to obtain a stable reading. Also, electrodes will require more time to equilibrate to temperature changes (see Section 5.4).

The time required to obtain a stable reading can be determined by recording pH versus time. Most often, the final pH reading is approached asymptotically, which allows the operator to see that a stable reading has been obtained or to predict the final reading (see Figure 2.3). The recording also provides a permanent record of response and pH values.

Either the glass or reference electrode can cause slow response. In order to determine which electrode is the source of the problem, a preequilibration test can be performed. Essentially, the test consists of preequilibrating one electrode to eliminate its response time, and then recording pH versus time when the other electrode is introduced into the sample (see Figure 5.2).

If a combination electrode is being used, another combination or glass and reference electrodes are employed in order to check both portions of the combination electrode in question. This is accomplished by connecting only that portion of the combination cable from the electrode that is being tested, and using the auxiliary electrodes to complete the pair; that is, the ferrule is connected to test the pH glass, and the pin jack is connected to test the reference. When testing the pH electrode of the combination, the pin jack from the reference of the combination should be connected to the solution ground to provide shielding for the glass electrode.

If a slow response is observed after preequilibrating the reference electrode, this may be caused by the glass electrode. Conversely, if a slow response is observed after preequilibrating the glass electrode, it may be caused by the reference electrode. The preequilibration time depends on how much time is required to obtain a stable reading. For example, if 10 minutes is required to establish a stable reading, then the glass should be preequilibrated for 10 minutes before introducing the reference electrode into the sample for its response test.

If the preequilibration response test shows slow response for the glass electrode after the reference electrode has been preequili-

GLASS ELECTRODE TEST

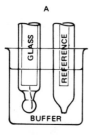
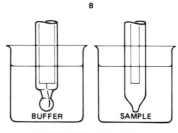
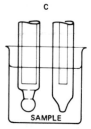

A	B	C
STABILIZE IN pH BUFFER	ALLOW REFERENCE ELECTRODE TO PREEQUILIBRATE IN SAMPLE	RECORD pH VALUE WITH TIME

REFERENCE ELECTRODE TEST

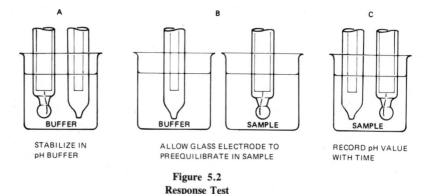

A	B	C
STABILIZE IN pH BUFFER	ALLOW GLASS ELECTRODE TO PREEQUILIBRATE IN SAMPLE	RECORD pH VALUE WITH TIME

Figure 5.2
Response Test

brated, it may be the result of a sample coating or an oil film on the bulb. Body oils transferred by touching the glass bulb with the fingers, for example, can cause a bulb to respond slowly.

Rejuvenation of the glass bulb will often provide faster response. This is accomplished by soaking the bulb in a solvent which will remove the coating or film. Any solvent may be used as long as it is also removed and the last soaking is in an aqueous–acidic solution. The choice of solvent or rejuvenating solution would depend on the most likely contaminate. In other words, in what samples has the glass bulb been immersed and in what solvents is that sample soluble? Most often an acid–base cycle soaking is sufficient to rejuvenate the glass response (see Figure 3.3). This cycle consists of soaking the bulb in 0.1 N HCl for 5 minutes, followed by 5 minutes in 0.1 N NaOH, and finally returning it to the acid for 5 more minutes.

If the preequilibration response test shows slow response for the reference electrode after the glass electrode has been preequilibrated, it may be the result of a clogged junction or an inappropriate junction material for the sample. A description of testing and clearing a high resistance or clogged junction is provided in Chapter 3, along with a description of the different types of junctions and their characteristics. A particular type of junction may exhibit slow response for a particular sample while another type does not. For example, a linen fiber junction is slow to respond in tris buffer while a ceramic junction provides much faster response.

5.4   TEMPERATURE

As discussed in Chapter 1, temperature has two effects on the pH measurement. Both the effect on the electrode potentials and on the ionic activity must be considered in order to apply proper measurement technique. How the temperature compensator of a pH meter functions to correct for changes in slope due to changes in temperature has already been discussed in Chapter 2.

Proper technique requires observation of the effect that temperature has on the measurement and the magnitude of error possible. First, the response of the electrodes is involved when temperature changes occur. In other words, do not standardize or take a reading until a stable reading is obtained. This may require considerable time if the electrodes must equilibrate to a large temperature change. Each potential involved with the electrodes (see Figure 1.3) is temperature dependent and must be established before a stable reading is obtained.

If the electrodes are going to a lower temperature, not only may the response require time, but the reading may be noisy. The glass resistance nearly doubles for every 7°C the temperature is lowered; that is, a glass bulb of 100 megohms at 28°C is about 1600 megohms at 0°C. If this electrode is in a high resistance sample, the noise and the sensitivity of the pH meter to the operator's movement are greatly increased.

The magnitude of error due to incorrect temperature compensation is of the order of 0.003 pH/°C/pH unit from standardization. For example, if the electrodes were standardized at pH 7 at 25°C, and a sample was measured at pH 4 but was at 23°C and the temperature compensator was not changed, the error would be 0.018 pH unit (0.003 × 2°C × 3 units). It is wise to keep in mind the magnitude of error possible by incorrect compensation when looking for sources of problems.

Also, for proper technique, the effect that temperature has on hydrogen ion activity should be considered. The label on a bottle of buffer lists the standardization values at specific temperatures for that buffer (see Table 4.1). The operator should always observe the buffer temperature and corresponding standardization value before calibrating the meter. For the most accurate results, the buffer and the sample should have as nearly equal a pH as possible and be brought to the same temperature. This minimizes any errors arising from differences between the ideal temperature-dependent slope factor and the actual electrode response due to temperature changes.

Remember that a solution has a specific pH value at all temperatures and that the pH of a sample should not be stated without stating the temperature. Often the pH value of a solution may be taken at 40°C, for example, but be desired in terms of 25°C. There is no method for calculating the pH changes due to this temperature difference unless the sample is studied the way NBS buffer solutions are studied to determine the effects that temperature has on the activity of the hydrogen ion in that particular matrix.

Also remember that the temperature compensator setting becomes more critical as the pH value deviates from 7. As discussed in Chapter 2, a meter is designed with zero millivolts at pH 7 and the temperature compensator control has no effect at this point. Since this point is the isopotential point of the electrode, any change in slope (mV/pH) due to temperature change has no effect.

As the pH value deviates from 7, however, the slope factor is multiplied by the number of units from the isopotential point, and thus has a large effect at the extremes of the pH scale (see Figure 1.6).

5.5 SEQUENCE OF OPERATION

The actual sequential steps involved in taking a pH measurement are simple. They may be a source of problems, however, since the steps include all possible sources of error previously discussed. The main considerations for establishing proper technique are to ensure proper working and noncontaminating electrodes, and to require stable readings before recording the pH value.

Standardization steps should include:

(a) Measure the temperature of the standard buffer solution and find its pH value at that temperature. Set the pH meter temperature compensator to the measured temperature.

(b) Rinse the electrodes to be used with a portion of the buffer or with distilled water. If water is used as the rinse, blot (do not wipe) the electrodes with a laboratory tissue to remove excess liquid. No blotting is necessary if a portion of the buffer is used as the rinse.

(c) Place the electrodes in a fresh portion of the buffer and activate the meter. Allow the electrodes to equilibrate with the buffer before setting the meter readout to the standardization value for that temperature.

The frequency of standardization will depend on the sample, the electrodes, and the desired accuracy. For example, a relatively clean sample measured with low potential drift electrodes would require standardization once a day if only 0.1 pH accuracy is required. If the standardization is a stable, reliable pH value, and the electrode exhibits low potential drift (<0.01 pH/hour) in buffers, there is no need for more frequent restandardization. In fact, what often occurs is that restandardization is a source of error because the meter is set to a value while the electrode is still influenced by the previous sample, and a complete stable equilibrium with the buffer has not been obtained. This is called a hysteresis effect.

If a higher degree of accuracy is required or the sample tends to coat or adhere to the glass bulb, as might be encountered in nonaqueous measurements, more frequent standardization is justified. The actual frequency should be determined empirically. In other words, by finding the drift rate of the electrodes in buffers, by stating a required accuracy value, and by determining the effect the sample has on the glass bulb and reference junction, a frequency of standardization can be established.

Sample measurement steps include:

(a) Place the meter in stand-by, remove the electrodes from the buffer, and rinse with an aliquot of sample or distilled water.

(b) Measure the sample temperature and set the pH meter temperature compensator to that value.

(c) Place the rinsed electrodes in the sample and activate the meter. Allow the reading to stabilize before recording the pH value.

5.6 OPERATIONAL PRECAUTIONS

There are a number of precautions an operator should observe in order not to damage the electrodes or pH meter and in order not to affect unknowingly the results of the pH measurement.

Protection of the equipment is maximized with the following precautions:

(a) The operator should be familiar with the functions of a pH meter and the electrodes. If this is a relatively new experience, knowledge can be greatly increased by reading the instructions supplied by the manufacturer of the pH meter and electrodes.

(b) Place the instrument in standby when the electrodes are not in solution. This protects the analog meter needle from suddenly colliding with the stop as it goes off scale. Also, if the electrodes are wiped while the readout is activated, the input amplifier can become saturated causing considerable delay before the reading will stabilize. The standby function is designed to separate the electrodes and protect the input amplifier.

(c) The pH glass electrode is, of course, subject to breakage. A stop on the rod (such as tape wrapped about the rod), which stops the electrode holder, can help prevent breakage of this electrode on the bottom of the beaker by preventing the electrodes from being lowered too far. A periodic word on the care in handling electrodes may save an electrode replacement cost.

(d) A plastic body electrode (polypropylene) should not be placed in boiling water; in fact, a boiling water procedure for clearing a reference junction should be used only after all other methods have failed to clear the junction (see Chapter 3).

(e) Electrodes should be used only with high impedance circuits. If current is allowed to flow, the electrodes can be permanently damaged.

The validity of results can be maximized with the following precautions:

(a) As discussed in Section 5.1, the electrodes should be blotted and not wiped. If the glass electrode does obtain a capacitance charge from the operator, a period of up to 15 minutes may be required to discharge the electrode. In the meantime, pH readings will be unstable and should not be considered valid.

(b) Periodic standardization of electrodes, as discussed in Section 5.5, should be observed. The frequency of this procedure should be determined empirically and related to the desired accuracy for a particular sample.

(c) If a glass electrode whose composition is designed for use in solutions of up to pH 11 is used above pH 11, considerable error will result. In order to compensate for this error, a correction factor can be added to the reading obtained. The correction factor is determined from a nomograph or chart supplied with the electrode. The correction factor is not as accurate as using a glass electrode designed for the full pH range. Therefore, if repeated measurements are to be made in the 11 to 14 pH range, a full-range glass should be used.

(d)   The electrodes should be presoaked before use. This provides a stable and fast-responding electrode. The hydrated layer on the glass bulb is established and the reference junction remains clear and unclogged by keeping these electrodes soaking (see Section 3.5).

(e)   The electrodes and the meter should be shielded against electrical or magnetic noise. If a pump motor, for example, sitting next to the meter and the electrodes is turned on, a noise signal will be observed on the display or recorder. Normally, this noise can be eliminated by moving the source away from the meter or the electrodes, or by proper grounding of the instrument and source. As discussed in Chapter 3, the shielding of the electrodes must be complete. Most often, the shielding on a glass electrode extends almost to the glass bulb. The sample solution should be above this point in order to complete the shielding. On a combination electrode, the reference filling solution provides the shielding for much of the electrode, while the shielding material provides protection for the top portion. Therefore, it is important to keep the reference filling solution level high to complete the shielding (Figure 3.14).

5.7   FACTORS OF ACCURACY

pH measurements are only as accurate as the equipment or solutions used and the technique employed. The magnitude of possible error from the components of these factors differs greatly.

A more complete investigation of these components has been discussed previously.

Table 5.1 summarizes the significant sources of error and their possible magnitude. When the meter malfunctions, the error is usually obvious. Normally, a fresh standard buffer solution is not a source of significant error. A glass electrode, which is short in span, can be mostly compensated for by a slope adjustment. If these components of accuracy are eliminated as being normally insignificant, the reference electrode and the components of proper technique remain as the probable cause of 75% of significant error or problems. The reference liquid junction is the most likely to be the cause of instability and slow response (see Figure 3.8).

These two components of accuracy are the areas in which the greatest precaution and thought should be applied. Careful selection and proper storage of the reference electrode are imperative in order to obtain a high degree of accuracy. Assuring that a stable reading is obtained before standardization or recording the pH value, blotting instead of wiping the electrodes, and using a standard buffer solution whose pH value is close to that of the sample are all components of high accuracy through proper technique.

TABLE 5.1

Factors of Accuracy

Factor	Component	Typical error
Equipment	Meter	±0.05 to ±0.001 pH, depending on meter
	Electrodes	
	Glass	±0.02 pH/pH unit from standard
	Reference	Junction potential may be large; depends on sample, junction material, and condition
Standard buffer		±0.005 to ±0.01 pH
Technique	Temperature	±0.003 pH/°C/pH unit from standard
	Response	
	Rinsing	} May be large; depends on operator
	Standardization point	

5.8   RECOMMENDATIONS FOR ACCURATE pH MEASUREMENTS

There are a number of techniques or precautions which should be employed if a high degree of accuracy is required. Some of the more important parameters are:

(a)   A standard *buffer* with a pH value close to the sample pH value should be used. If possible, one of the primary buffers (4.01, 6.86, 9.18 should be employed. A buffer which is fresh and precisely prepared will provide increased accuracy.

(b)   The electrodes selected for the pH measurement should be *tested* and found to meet their performance specifications. The reference electrode whose junction exhibits the least junction potential and the fastest response in the sample should be selected.

(c)   A pH meter with *readability* of at least ±0.01 pH should be used. The meter should have a *slope* control to adjust the span for nonideal electrodes if sample pH values vary over a wide pH range.

(d)   Sufficient stabilization *time* should be allowed for each measurement. This is particularly important when a temperature change is involved or a large junction potential is developed. The time required to obtain a stable reading can best be observed by recording the pH values versus time.

(e)   Sample and buffer *temperatures* should be the same. If long-term, accurate measurements are involved, a water bath should be employed.

Chapter 6

Applications

The applications of pH measurements are too numerous to discuss separately. The measurement of pH is used in producing the ink on this page, the color in a shirt, the shine on a car bumper, and is vital to each of us in our bodies. The most reasonable approach to a discussion on pH applications is to discuss a general approach and then illustrate it with examples of some common difficult samples.

6.1 GENERAL APPROACH

There are four common sources of error that can occur when making pH measurements on difficult samples. The *first* is high sample resistance which can result in slow response and increased noise pickup. The *second* is lack of compatibility of the reference filling solution with the sample or poor performance from a particular type of junction in the sample. This lack of compatibility results in a large unstable liquid junction potential which can cause slow response, instability, and/or significant error. The *third* common source of error is contamination of the sample. This may be the result of gas or liquid contamination; for example, absorption of carbon dioxide into distilled water, carry-over of buffer on electrodes into a low ionic strength sample, or retention of a sticky, viscous sample on the glass bulb. The *final* common source of error is the sample itself. Because of unusual conditions such as high or low temperature or pressure, or because the sample is a dry solid, perhaps with a flat surface, it becomes difficult to measure the pH. Table 6.1 shows that one or more of these sources of error can be present for many difficult samples.

For each common source of error, there is a suggested method to follow in order to minimize the error. These procedures are outlined in Table 6.2.

The possible procedures which may be employed to reduce a *contamination* error depend on the source. A viscous sample which sticks to the glass bulb may require multiple solvent rinses; that is, rinsing with a solvent which dissolves the sample. In order to prevent a gas contamination, the sample must be protected from the contaminant by blanketing the sample with an inert gas. To remove a gas contaminant from the sample prior to the pH measurement, the sample can be purged with an inert gas.

Adding a neutral salt to the sample is one procedure for reducing error in a *high resistance sample*. As previously discussed, this changes the ionic strength slightly and therefore the hydrogen ion

TABLE 6.1

Sources of Error

	Error sources			
Difficult samples	Contamination	Sample resistance	Junction potential	Sample conditions
Nonaqueous oils	×	×	×	
Distilled water	×	×	×	
High salt			×	
Solids (Flat, Dry)		×	×	×
Viscous Slurry	×	×	×	
Extremes of temperature or pressure				×
Strong acid or base			×	×

TABLE 6.2

Minimizing Error

Source	Possible procedure
Contamination	Multiple rinse with solvent or aliquot of sample
	Purging or blanketing sample with inert gas
High sample resistance	Add a neutral salt to the sample
	Connect sample or a surrounding shield to solution ground terminal
Large liquid junction potential	Alter reference filling solution
	Use an auxiliary salt bridge with an intermediate electrolyte filling solution
	Use a reference electrode with different type of junction
Sample conditions	Add distilled water to solid sample
	Use flat bulb combination electrode
	Use pressurized or solid state reference electrode
	Sterilize electrodes

activity. This usually introduces an insignificant error, however, when compared to the error which may arise without the addition of the salt. Connecting the sample or the shield surrounding the sample to a solution ground terminal may be necessary in high-resistance sample measurements when the sample container is isolated from earth ground and has stray ac noise.

The three possible procedures for reducing the *liquid junction potential* error are directed at providing greater compatibility between reference and sample. This is accomplished by altering the filling solution within a reference electrode body, providing an intermediate electrolyte in an auxiliary salt bridge, or changing the type of junction used in making the measurement. These procedures are discussed in Chapter 3 as well as in the examples given in Section 6.2.

The possible procedures to follow in order to reduce the error which may result from special *sample conditions* are numerous and depend on the type of condition. For example, a solid sample may require the addition of water or emulsification in a blender. A flat surface gel sample may require a flat bulb combination electrode in order to make contact with both the glass membrane and the liquid junction at the same time. A high pressure sample may require the use of a solid state reference or a reference electrode with a side arm for pressurization.

6.2 DIFFICULT SAMPLES

The discussions of difficult sample types incorporate the methods for minimizing error and suggest procedures for making measurements in each type of sample.

6.2.1 Nonaqueous pH Measurements

There are several difficulties which are inherent with pH measurements in mixed solvents. The objective of this section is to explain briefly the concept of pH in nonaqueous solvents and to discuss minimization of the difficulties.

Hydrogen ion activity, to which the glass electrode responds, is affected by the medium (solvent) in which it is contained. pH scales based on the hydrogen ion activity (a_{H^+}) are limited to a single solvent or solvent mixture. Comparison of a_{H^+} in an aqueous buffer with that of a nonaqueous solvent, therefore, does not have quantitative or thermodynamic significance. In order to obtain some useful results, however, an operational pH formula is most often adopted when the measurement in the nonaqueous solvent is compared with that in an aqueous solution. If the nonaqueous pH measurement is stable and can be correlated to some results, the absolute hydrogen ion activity need not be known. The relative pH value can be used as an indicator to alter the process or proceed in some corrective manner if the pH value changes dramatically.

Added to the medium effect on activity is the hindering of the pH glass functioning by the solvent dehydrating the glass, by high sample resistance, and by large liquid junction potential developed at the reference electrode. These factors make nonaqueous pH difficult to measure and interpret.

A. Solvents

The medium effect reflects the electrostatic and chemical inter-actions between the ion and the solvent, of which the primary interaction is solvation. This effect influences the ionic activity and can be related by comparing the standard free energy in a nonaque-ous solvent with that in water. For example, the activity of hydrogen ions is much greater in ethanol ($\approx$200 times) than in water.

Although the medium effect is not wholly an electrostatic quan-tity, one method of estimating this effect is through the dielectric constant. The lower the dielectric constant of the solvent, the greater the associating power the ions have for each other. Conse-quently, there is a great tendency for such ions to aggregate into inactive pairs or even larger complexes. A solvent of high dielectric constant will generally serve as a good solvent for ionic com-pounds. In a high dielectric solvent such as water, the ion pairs are almost completely dissociated into free ions (see Table A.2).

The low ionic strength and low conductivity of some nonaqueous solvents (see Table A.4) may result in severe noise pickup and large liquid junction potentials. These effects can be minimized by increasing the ionic strength of the solvent with a neutral electro-lyte such as a quaternary ammonium salt. The addition of a neutral salt to the solvent increases its ionic strength, however, and consequently affects the hydrogen ion activity. Normally this effect is insignificant when compared with the potential error without the salt.

A supporting electrolyte that produces negligible alkaline error, such as salts of magnesium, calcium, barium, or organic cations, should be used. Lithium chloride or sodium perchlorate are recom-mended for alcoholic media. Some common solvents in which tetrabutylammonium iodide (Bu_4NI) and tetraethylammonium perchlorate (Et_4NClO_4) are soluble are listed in Chapter 3.

B. Solvent Properties

In order to understand better the effect that different solvents will have on the pH, a brief discussion of the characteristics and types of solvents is necessary.

The pH of samples in a nonaqueous solvent is greatly influenced by the acid–base properties of the solvent and the potential which is developed at a pH glass bulb.

Nonaqueous solvents may be grouped on the basis of their acid-base properties (Table 6.3).

Amphiprotic

This type of solvent has upper and lower pH limits. The range is determined by the acidic and basic character of the solvent. The length of the pH scale in each medium, in pH units, is equal to $-\log K_s$, where K_s is the autoprotolysis constant of the solvent. The pH at the acidic end of the scale in each solvent is $-\log f^m_{H^+}$. The solvent acts as both a proton donor and as a proton acceptor. Examples are water and alcohol.

Acidic or Protogenic

The acid strength of various solvents differs according to the ability of the chemical structure of the solvent to bind the proton. The highest pH obtainable with this type of solvent is limited by its acid character even though a strong base is added. There is no low pH limit, however. The solvent is a proton donor only and never acts as a proton acceptor.

TABLE 6.3

Nonaqueous Solvents

Type	Description	Examples
Amphiprotic	Both acid and basic properties	Alcohol, water
Protogenic	Acid properties	Phenols, carboxylic acids
Protophilic	Basic properties	Dimethylformamide
		Pyridine
		Ethylenediamine
		Liquid ammonia
Aprotic (inert)	Neutral	Benzene, Chloroform
		Acetonitrile
		Dioxane, hydrocarbons

Basic or Protophilic

This type of solvent is a proton acceptor, and the lowest pH limit is determined by the tendency of the proton to escape from the solvent. There is no upper pH limit since proton activity gradually decreases.

Aprotic

This type of solvent cannot exist as a proton donor or a proton acceptor and therefore does not have acidic or basic pH limits (examples are hydrocarbons). pH limits for different types of solvents are shown in Figure 6.1.

A mixture of solvents, to take advantage of the favorable characteristics of each, can reduce the useful potential range. For example, as methanol is added to pyridine the acidic limit of pyridine remains fairly constant while the basic limit is reduced.

C.   Glass Electrodes in Nonaqueous Solvents

The ability of an electrode to respond to changes in pH is associated with the water content of the glass. The glass bulb swells slightly when immersed in a solution, and a hydrated layer is formed as the water penetrates into the silicate network. This layer seems to facilitate the movement of ions in the glass and to lower the electrical resistance.

If the electrode is allowed to dry by nonaqueous dehydration, it loses its pH function. Therefore, it is imperative that a glass electrode used for nonaqueous pH measurements be soaked periodically in water to rejuvenate it. Even with partial dehydration, however, glass electrodes function properly for a moderate time period in nonaqueous solutions that have a dielectric constant as low as 2.3 (see Table A.2).

In nonaqueous solvents, the ion-exchange equilibrium potential between the hydrogen ions in the sample and the ions in the glass bulb is established slowly. When a solution is changed from basic to acidic or visa versa, a slow response is observed. One reason for the slow response is the lower level of hydration.

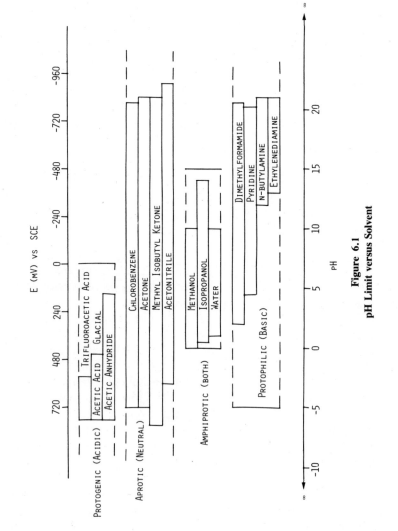

Figure 6.1
pH Limit versus Solvent

Other factors influence the response time, such as buffer capacity of the solvent and the type of glass membrane. Normally, compared with full-range pH glass, a glass with limited pH range will offer the better response because of its lower resistance and more rapid equilibrium of the hydrated layer.

One reason for the observation of drift when making a nonaqueous pH measurement is that the thickness of the gel layer, which surrounds a glass bulb, is changing due to dehydration. When the hydration rate equals the dissolution rate of the outer glass layer, equilibrium is established.

One method to limit the amount of hydration of a glass bulb in order to decrease response time has been to etch the glass bulb and then to hydrate it in strong acid. The etching of the bulb is done in 2% hydrofluoric acid in water for 2 minutes at room temperature; it is then hydrated in dilute HCl for about 1 hour. This partially hydrated electrode is not suitable for use in aqueous solutions because further hydration will cause drift. Further hydration can be limited by storage in the organic solvent to be used, but the electrode must be immersed in water for a few minutes before use. This etching procedure does limit the electrode life and is normally not performed when response time is not critical.

D. Buffers

Most often an aqueous pH buffer solution is used to standardize the pH measuring system. If the measurement is to be made on a nonaqueous sample, the correlation between hydrogen ion activity in an aqueous standard and in a nonaqueous sample is not valid. Nonaqueous buffers which provide a more realistic pH value in nonaqueous samples through standardization under conditions of similar medium effect and liquid junction potential are described in Chapter 4.

E. Reference Electrodes

When measuring the pH of a sample in mixed solvents or nonaqueous solvents, a large liquid junction potential is developed. This may result in an unstable reading or require a long time for stabilization. The junction potential is developed because of the different rates of interdiffusion of ions in the nonaqueous solvent compared with the aqueous filling solution. Examples of the magnitude of pH error which can be the result of a liquid junction potential are listed in Table 6.4.

Several steps can be taken to minimize the junction potential. These steps include selection of an appropriate type of junction, providing a compatible filling solution, and/or separating the reference from the sample by the use of an auxiliary salt bridge.

Type of Liquid Junction

A junction with minimal flow rate is often preferred in order to reduce possible salt error. This is particularly true when alternate electrolytes are used. The quartz junction has a slow flow rate and provides good stability in nonaqueous samples.

Filling Solutions

A normal calomel reference electrode with aqueous filling solution may be used in nonaqueous solutions, but most often develops a large or unstable junction potential. One means of reducing this potential in any nonaqueous solution is by changing the filling solution so that it is more compatible with the solvent. For example, a methanol solvent saturated with potassium chloride may provide a more stable reference electrode.

Another example of an alternate filling solution is the use of 90% glacial acetic acid plus 10% saturated aqueous lithium chloride for

TABLE 6.4

Liquid Junction Error

Ethanol (wt %)	pH Error due to liquid Junction
0	0
20	−0.02
35	0.11
50	0.42
65	0.74
80	1.26
100	2.35

TABLE 6.5

Reference versus SCE Standard Potentials

Solvent	Electrode	Millivolts
Methanol		
20%	Calomel[a]	255
43%	Calomel	242
68%	Calomel	217
99%	Calomel	103
Dioxane		
20%	Calomel	250
45%	Calomel	210
70%	Calomel	113
82%	Calomel	−1
Ethylene glycol		
19%	Calomel	257
50%	Calomel	236
78%	Calomel	201
Acetic acid	Ag/AgCl, KCl	230
	Ag/AgNO$_3$	870
	Calomel	270
	Hg/Hg$_2$SO$_4$, K$_2$SO$_4$	690
2,4-Lutidine	Calomel	330
	Hg/Hg$_2$SO$_4$, K$_2$SO$_4$	290
2,6-Lutidine	Calomel	450
	Hg/Hg$_2$SO$_4$, K$_2$SO$_4$	360
2-Picoline	Calomel	420
	Hg/Hg$_2$SO$_4$, K$_2$SO$_4$	390
Pyridine	Hg/Hg$_2$SO$_4$, K$_2$SO$_4$	340
Quinoline	Ag/AgCl, KCl	170

[a] Calomel is Hg/Hg$_2$Cl$_2$, KCl.

measurements in glacial acetic acid with acetic anhydride added. Since the electrolyte contains chloride, it may be placed directly in a reference electrode salt bridge that normally contains potassium chloride. Once the reference potential has been established with a new filling solution, the reference electrode becomes a special reference electrode dedicated to nonaqueous measurements. Some other reference electrodes are listed in Table 6.5 with their standard potential versus SCE at 25°C. The conditions required to provide effective filling solutions are discussed in Chapter 3.

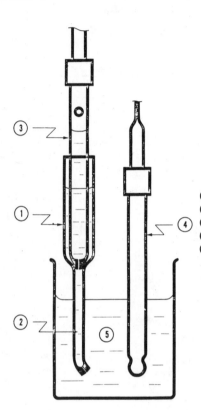

(1)   Salt bridge
(2)   Intermediate electrolyte filling solution
(3)   Reference electrode
(4)   Glass electrode
(5)   Sample

Figure 6.2
Nonaqueous pH Measurement

Salt Bridge

The purpose of a salt bridge built as part of the reference electrode is to provide contact with the sample while surrounding the reference internal with a known electrolyte. This filling solution in contact with the internal establishes the reference potential. If the filling solution is altered, the potential changes. There are several conditions which may require the use of an auxiliary salt bridge. This auxiliary salt bridge, which is a glass body with a liquid junction at one end as shown in Figure 6.2, provides separation between the reference electrode and the sample. Reference electrodes which have a double junction feature built in are also available. Thus, the reference potential is maintained while meeting the conditions required of an alternate filling solution.

The use of an auxiliary salt bridge is required, for example, to prevent contamination of the sample by the potassium chloride filling solution, or if an electrolyte having other than a chloride anion is being used with a calomel internal. Thus, potassium chloride may be used in the reference electrode to provide the stable reference potential, while a second filling solution is used in the auxiliary salt bridge. This type of setup shown in Figure 6.2 may also be required to provide greater stability.

An example of using an auxiliary salt bridge in making pH measurements in dimethyl sulfoxide (DMSO) would be a reference electrode containing a methanolic–KCl filling solution placed in an intermediate solvent–electrolyte consisting of 50% DMSO and 50% methanol, with a small amount of tetraethylammonium perchlorate. This intermediate solution serves as a bridge between the reference and sample solutions. The liquid junction potential is reduced by the filling solutions which provide solution steps that are more compatible than aqueous to DMSO.

In general, a more precise selection of solvent and electrolyte which compose an intermediate filling solution can reduce the liquid junction potential.

F. pH Meters

Nonaqueous samples add significant resistance to the circuit. They are often at the level of the pH glass bulb resistance or greater. When the high resistance values of the circuit are added together, it becomes obvious why a noisy reading is often observed on a pH meter when the electrodes are immersed in a nonaqueous sample. In fact, if this resistance is the same magnitude as the input resistance of the meter, a voltage division occurs, and a pH span error results in reduced pH sensitivity for the meter. This resistance is the reason for the suggested addition of electrolyte to the sample in Section A.

The choice of pH meter can also help provide better stability when making pH measurements in nonaqueous solvents. A meter which has a bias current of less than 5 picoamps will help to provide this stability.

Also discussed in Section B is the fact that greater than the normal 0–14 pH range may be encountered. Many digital meters are capable of displaying -19.99 to $+19.99$ pH. In other meters, such as the analog meters, the scale is limited. The millivolt range of most pH meters provides greater range, such as ± 1400 mV, than the pH range 0–14 which is ± 421 mV at 25°C. Therefore, the pH value can be calculated by comparing the millivolt values obtained in standards with the millivolt value obtained in a sample. Of course, there is reduced readability with the larger millivolt range. The equation for comparing pH values as seen by the pH meter is

$$pH_x = pH_s - \frac{E_x - E_s}{S}$$

where

pH_x is the pH of the sample (x),
pH_s the pH of the standard (s),
E_x the millivolt value of the sample as read by the pH meter,
E_s the millivolt value of the standard as read by the pH meter, and
S the slope, normally 59.16 mV/pH unit at 25°C.

Example

Sample $pH_x = ?$, $E_x = 240$ mV, $S = 60$ at 30°C.

Standard pH 4.01, $E_s = 180$ mV.

$$pH_x = 4.01 - \frac{240 - 180}{60}$$

$$= 4.01 - \frac{60}{60}$$

$$= 3.01$$

G. *Recommendations for Nonaqueous pH Measurements*

There are a number of precautions which involve equipment, sample preparation, and measuring technique to provide more accurate results in nonaqueous solvents.

Equipment

(a) Use a low-resistance general-prpose pH glass electrode for its small ion-exchange capacity.
(b) Use a slow-flowing reference electrode such as a quartz junction calomel electrode.
(c) Use a pH meter with a low bias current and high input resistance.
(d) In the reference electrode, use a filling solution which is compatible with the sample.
(e) Use an auxiliary salt bridge when necessary to incorporate an intermediate electrolyte which is compatible with the sample.

Sample Preparation

(a) Add a small amount of neutral electrolyte to reduce the sample resistance. Quaternary ammonium salts are often used.
(b) The added electrolyte should contain ions which produce little alkaline error.

Technique

(a) Use a nonaqueous buffer when possible.
(b) Between successive measurements, the electrodes should be rinsed with the nonaqueous solvent used to dissolve the sample.
(c) If an aqueous buffer is used, the glass and reference electrode should be soaked in the nonaqueous solvent for 10 minutes after standardization and before use in the sample. Etching of the glass bulb may be employed to reduce response time.
(d) If the reading begins to drift after considerable time in a

- nonaqueous solvent, the pH glass bulb must be hydrated by immersion in aqueous buffer.
(e) Storage solutions for electrodes depend on the length of storage time. For short-term storage, use the solvent, and for long-term storage, use aqueous pH 4 buffer.
(f) Maintain buffer and sample at the same temperature.
(g) Allow sufficient stabilization time for each measurement before taking a reading.

References

1. Karlberg, B., Response-Time Properties of Some Hydrogen Ion-Selective Glass Electrodes in Nonaqueous Solutions, *Anal. Chem. Acta* **66**, 93 (1973).
2. Farinato, R. S., Tomkins, R. P. T., and Turner, P. J., A Study of pH Glass Electrode Drift in Acetonitrile Buffer Solutions, *Anal. Chem. Acta* **70**, 245 (1974).
3. Popovych, O., Estimation of Medium Effects for Single Ions and Their Role in the Interpretation of Nonaqueous pH, *Anal. Chem.* **38**, 558 (1966).
4. Ritchie, C. D., and Heffley, P. O., Acidity in Nonaqueous Solvents, I, Picolinium Ions in Methanol, *J. Am. Chem. Soc.* **87**, 5402 (1965).
5. Kucharsky, J., and Safarik, J., "Titrations in Nonaqueous Solvents." Elsevier, New York, 1965.
6. Sisler, H. H., "Chemistry in Nonaqueous Solvents." Van Nostrand-Reinhold, Princeton, New Jersey, 1961.
7. Kucharsky, J., and Safarik, J., "Titrations in Nonaqueous Solvents." Elsevier, New York, 1965.
8. Audrieth, L. F., and Kleinberg, J., "Nonaqueous Solvents." Wiley, New York, 1953.
9. Waddington, T. C., "Nonaqueous Solvent Systems." Academic Press, New York, 1965.
10. Huber, W., "Titrations in Nonaqueous Solvents." Academic Press, New York, 1967.
11. Gyenes, I., "Titrations in Nonaqueous Media." Van Nostrand-Reinhold, Princeton, New Jersey, 1967.

6.2.2 Dry, Porous, or Flat Solid Samples

By definition, the pH of a substance requires a liquid. Also, measurement with electrodes requires conductivity between the glass membrane and the reference junctions. Therefore, the only means of obtaining the pH of a solid dry substance is by adding liquid. If the sample can be prepared as finely divided particles, greater surface area is exposed to the liquid, and thus the extraction of ions from the substance into the liquid is more efficient than if the liquid is placed on a flat surface. The flat-surface measurement, however, may be required if the sample cannot be destroyed, such as in the testing of the pH of ancient documents.

There are two basic methods for measuring the pH of a dry solid. The first requires about 1-hour contact of 20 to 50 ml of distilled water with 1 gram of finely divided sample. The sample is not filtered and the pH measurement is taken on the supernatant liquid. The other method is to take a pH reading after one or more drops of distilled water is placed on the surface of the sample. This measurement requires a flat bulb combination electrode to be placed on a wetted area sufficiently large to provide contact between the glass membrane and reference junction.

In order to obtain repeatable pH measurements which are representative of the dry substance, four factors must be investigated. Two factors are similar to the characteristics of buffers; that is, the buffering capacity and the dilution factor must be investigated. If the sample has little buffering capacity, carbon dioxide pickup from the atmosphere will produce significant error. For a flat-surface measurement in which the small amount of water used is spread over a large area, the CO_2 pickup is quite significant. The use of a blanket of inert gas or taking the measurement after a set elapsed time may reduce this error. Also, a good-quality deionized water, boiled for 10 minutes and cooled while excluding atmospheric CO_2, could be used as the water added to a dry surface.

The dilution factor should be investigated to determine the effect on the pH of different volumes of distilled water added to the 1-gram sample. This parameter will dictate how important the exact volume of water added to the sample is to the accuracy of the measurement.

The two other factors which should be investigated are the extraction rate and the electrodes which will provide the most stable reading. If the pH variation with time is not due to CO_2 pickup or electrode instability, it is due to the extraction rate. If, for example, the added water penetrates the sample slowly, the concentration of extracted chemicals changes with time and thus affects the pH. Stirring is often used with a ground-up sample to decrease the extraction time, but the completion of extraction is largely dependent upon the nature of the sample.

The proper selection of electrodes is discussed in Chapter 3. The primary concern is to select a reference electrode which provides a stable liquid junction potential in the sample. With limited sample volume, the combination electrode is often required, but an electrode pair with a sleeve junction reference is frequently used in a colloidal sample because of its low junction resistance and ease of cleaning (see Figure 6.3).

Soil Samples

Soil is a dry solid sample whose pH is frequently measured. There are three recognized methods of analysis which follow the general procedures already outlined. The first procedure entails adding sufficient distilled water to the soil sample to make a thin paste and then allowing the mixture to stand for 5 minutes while extraction is completed. The second method incorporates a 1:1 soil-to-water ratio. This mixture is stirred at regular intervals for

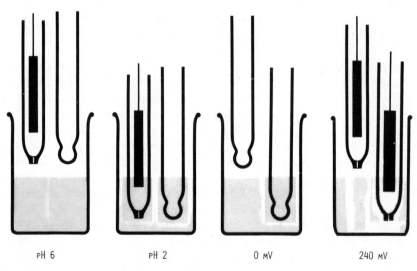

pH 6   pH 2   0 mV   240 mV

Figure 6.3
Liquid Junction Potentials from Suspension Effect

about an hour before the measurement is taken. Obviously more extraction of the ions in the soil will occur with the longer contact time. The third method is a combination of the first two, with a more exact mixture of a saturated soil paste standing 1 hour prior to measurement.

All three methods will give repeatable results, but each method will give a slightly different pH value for the sample. With this in mind, it becomes obvious that the preparation of a dry sample is important to the accuracy of the measurement.

6.2.3 Slurries, Sludges, Colloidal, or Viscous Samples

The problems which occur when measuring the pH of this sample type are most often associated with a liquid junction potential. Long electrode stabilization time, constant drift, or significant pH error are problems which may be encountered. For example, a slurry may settle and result in a two-layer solution of apparently different pH values. Figure 6.3 illustrates the liquid junction error by first measuring in the supernatant liquid and then in the sediment or colloidal portion of the sample, with two different pH values obtained. If two glass electrodes are used, one in the supernate liquid and one in the sediment, no potential difference is observed; that is, the glass electrodes are bucked against each other showing the same pH in both portions of the sample. If two reference electrodes are used, a potential difference is observed between the layers of the solution.

The methods of reducing the liquid junction potential have already been discussed in Chapter 3. Basically, they include proper selection of the junction or altering the filling solution. Most often, a sleeve junction reference electrode will afford the greatest stability and the least junction potential in this type of sample. It is also very easy to clean, which becomes important with a sticky-type sample that is apt to clog the normal fiber-type junction.

A change in the filling solution or the use of an auxiliary salt bridge may be required with such viscous samples as latex. A reaction of the sample with a potassium chloride filling solution can be avoided by using an auxiliary salt bridge with an alternate electrolyte filling solution.

6.2.4   Distilled or High Purity Water

There are many different sources of high purity water samples such as boiler feedwater and condensate streams. The pH of these solutions is often measured and controlled to minimize corrosion of boiler hardware. Too low a pH will result in excessive corrosion of iron components, while too high a pH will result in an attack on the copper-containing parts.

It is difficult to measure accurately the hydrogen ion activity of high purity water having a low conductivity such as less than 10 micromhos. The problem arises from the high resistance and unbuffered nature of high purity water and from the liquid junction potential that is developed; that is, the measurement is likely to be noisy because of the high sample resistance, it is likely to drift because of carbon dioxide adsorption, and it is likely to require considerable stabilization time or be in error because of a large liquid junction potential.

The noise observed from measuring low conductivity samples arises from the fact that the meter is more susceptible to stray ac fields and electrical interference when the combined sample and glass bulb resistance are in the circuit. Two methods, one directed at handling the sample resistance and the other at reducing the sample resistance, are used. First, a meter with low bias current which is able to handle high resistance between the inputs may be used. Secondly, a neutral electrolyte can be added as previously suggested for nonaqueous samples. If a neutral electrolyte is used, some accuracy is sacrificed since increasing the ionic strength affects the hydrogen ion activity. Only a small amount of electrolyte is needed to reduce sample resistance and can often be provided by a fast-flowing sleeve junction reference.

The measurement difficulty associated with absorption of carbon dioxide can be minimized by blanketing or purging the sample with an inert gas such as nitrogen or measuring the sample quickly after exposure to the atmosphere. If the sample is not stirred and the electrodes are placed deep in the solution, the concentration of dissolved carbon dioxide will require considerable time to reach equilibrium with the atmosphere. Thus, the pH measurement can be made before being affected by carbon dioxide.

The unbuffered nature of the sample makes it susceptible to large pH changes from any contaminant. For example, if the electrodes were not completely rinsed after standardization in a buffer, and a small amount of buffer were transferred to a high purity sample, this would greatly affect the sample pH value. In other words, a small amount of pH buffer at $0.1\,m$ ionic strength can greatly influence a large sample at $10^{-6}\,m$ ionic strength. Because of this susceptibility to contamination, special precautions must be taken to prevent it.

The measurement difficulty associated with the liquid junction stems from the large difference in ionic strength between the reference filling solution and the high purity sample. A concentration gradient can be developed at the junction. If the sample is stirred, a noisy measurement is observed because of changes in this gradient.

Another source of junction potential can arise when using a combination electrode. Although the combination electrode has an advantage in that the distance between the glass and the reference is small and constant, thus minimizing sample resistance path, it most often has an inherent disadvantage. The filling solution of a combination electrode with a silver chloride internal contains many complexes of silver chloride when at $4\,M$ KCl. If this filling solution is diluted, the complexes will precipitate. This can be shown by adding a drop of water to the filling solution and noting the formation of a milky white precipitate. If these complexes precipitate in the junction by contact with high purity samples, a large junction potential can result.

A calomel sleeve junction electrode can provide low junction potential while adding a small amount of potassium chloride electrolyte to the sample and thereby reducing its resistance. Some users reduce the concentration of the normally saturated KCl filling solution to $0.1\,M$ KCl in order to minimize the junction potential and to decrease stabilization time. This electrode potential is listed in Table 3.5.

6.2.5   High Salt Samples

At the opposite extreme from high purity samples are samples with high salt content. The main problem with measuring this type of sample is the liquid junction potential. As discussed in Section 3.2.4 the ions in the filling solution are at high concentration in order to provide the dominant effect on the junction potential. With the sample having a high concentration of competing salts, the filling solution electrolyte may not provide the dominant ions. If the electrodes are standardized in a 0.1 m or less ionic strength buffer, the liquid junction potential may be significantly different than in the sample.

One method of reducing this error is by the use of a high salt buffer. For example, a seawater buffer described in Chapter 4 provides greater accuracy when measuring the pH of seawater. Another method is the use of a fast-flowing low-resistance junction such as the sleeve junction.

Other users have slightly altered the filling solution if the high salt sample is at the pH extremes. Since the hydrogen and hydroxide ions have such a large limiting equivalent conductance value, the high salt sample at the pH extreme would be even more dominant over the filling solution electrolyte than a neutral high salt sample. Therefore, a slight amount of acid or base has been added to the filling solution to make it more compatible with the sample. This does not reduce the liquid junction potential a great deal, but it does decrease the stabilization time required in the samples.

6.2.6   Extremes of Temperature or Pressure

There is little that can be done to measure pH in addition to following the manufacturer's electrode specifications on temperature and pressure. Typically, these specifications for normal electrodes are −5°C to 100°C and up to 150 psig.

A.   Glass Electrodes

Since the resistance of the glass bulb approximately doubles for every 7°C decrease in temperature, the lower temperature limit of −5°C is stated only for glass electrodes with relatively low resistance. A slower response and noisier readings can be expected at lower temperatures. If the glass electrodes are used continuously at temperatures higher than approximately 80°C, the lifetime is considerably shortened. The hydrogen electrode not normally employed for practical pH measurements will perform above 100°C with reliable results.

B. Reference Electrodes

Another cause for the limits of the temperature range is the freezing or boiling of the reference filling solution. Methanol has been used in the filling solution to depress the freezing point, but the freezing point of aqueous saturated potassium chloride is −11°C, which is near the pH meter limit because of the high glass electrode resistance.

A reference with a calomel internal should not be used at temperatures of 80°C or higher for a prolonged period. The silver-silver chloride reference, however, is better suited for high temperature measurements. Often the reference electrode is placed in a remote, cooler position. Thermal diffusion potentials are created by the use of the remote salt bridge, but their influence is compensated for by buffer standardization under similar conditions.

Under conditions of high pressure, the positive flow of filling solution from the reference junction must be maintained. Otherwise the sample is forced up the junction and contaminates the internal. Two common methods of ensuring positive flow are the use of a reference with a side arm or the use of a solid state reference. The reference with a side arm can be pressurized with slightly greater pressure than the sample. Some solid state reference electrodes are constructed with a pressurized internal which prevents diffusion of the sample into the electrode body.

6.2.7 Acid–Base Titrations

The objective of this brief discussion on titrations is to relate the parameters discussed in other chapters to the problem associated with this technique. In other words, an aqueous or nonaqueous acid–base titration is nothing more than a series of pH measurements. All of the precautions, techniques, and equipment previously described apply. Other sources of information on acid–base titrations should be consulted for detailed procedures.

Titrations are generally considered to be among the most useful and accurate analytical techniques. Although a wide variety of systems is available from which the best system for a particular analysis can be chosen, the basic principle for all titrations is the same. An unknown quantity of sample is titrated to an end point with a known amount of titrant under conditions for which the reaction chemistry is known and reproducible. In all cases, the equivalence point or end point is characterized by a relatively sharp change in the hydrogen ion activity which is used to follow the course of the titration.

Potentiometric acid–base titration is one of the most accurate and most widely applicable methods. In the absence of interferences, the accuracy which can be achieved is usually limited only by volumetric errors, preparation of titrants of known strength, and factors related to equilibrium constants of the titration reaction. Problems involved in the availability and selection of an indicator are avoided. Also, since the property measured is a change in potential rather than a change in the absorption of light, colored and turbid solutions do not impose difficulties.

This technique involves plotting the pH indicated by the glass electrode versus the volume of the titrant. Normally, an S-shaped curve is obtained with the equivalence point characterized by a maximal value of the slope. The titrant may be added in large increments before the equivalence point is reached, in small equal increments through the equivalence point, and in large increments after the equivalence point is passed. After each increment is added, sufficient *time* should be allowed for the reaction to occur and for the electrode to reach a reasonably constant potential before taking a reading.

A. Types

Acid–base titrations usually fall into one of the following classifications:

(1) strong acids or bases,
(2) weak monoprotic acids or bases,

(3)   polyprotic acids or bases,
(4)   mixtures of weak acids or weak bases,
(5)   anions of weak acids and cations of weak bases (displacement titrations).

Examples   (see Figure 6.4)

An example of the first and third classes would be:

(A)   titration of hydrochloric acid with sodium hydroxide (type 1);
(B)   titration of phosphoric acid with sodium hydroxide (type 3).

Potentiometric titrations provide a high degree of accuracy and total concentration measurement. When making a direct pH measurement, only the free active hydrogen ion is sensed and the value obtained is relative to another solution, the pH buffer. With titrations, the calculated answer provides total hydrogen ion concentration provided the reaction between the titrant and the hydrogen ion has a sufficiently fast rate, is in one direction without

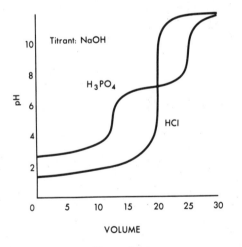

Figure 6.4
Acid–Base Titrations

side reactions, and is stoichiometric. The end point is determined relative to other potential measurements in the same solution. Therefore, the liquid junction potential is likely to remain constant, thus eliminating this source of error.

Nonaqueous titrations are frequently desirable or required because of the increased sensitivity, improved selectivity, or greater solubility achieved with nonaqueous solvents. A far greater number of acids and bases can be determined in nonaqueous solvents than in aqueous media. This is primarily true because of the numerous organic acids and bases that require organic solvents. Properties such as dissolving or solvating, diffusion or equilibrium constants, acidity or basicity, and dielectric constant or polarity extend the capability of titrimetry to a far wider range when nonaqueous solvents are used.

Water is amphiprotic by nature; that is, it can act as a base or an acid. However, it has a very limited range of acid and base strengths (see Figure 6.1). Strong acid in the presence of water (H_3O^+) is not sufficiently acidic to give a sharp end point for weak organic bases such as aniline. Similarly, a strong base is not sufficiently basic to give a sharp end point for weak organic acids such as phenol. The high polarity of water may interfere with the determinations of relative strengths of different acids or bases as shown in Figure 6.5.

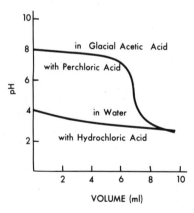

Figure 6.5
Nonaqueous versus Aqueous Acid–Base Titration: Potassium Acid Phthlate

B. Technique

In order to obtain high accuracy when performing acid–base titrations, proper technique must be observed. This includes those parameters discussed in Chapter 5, such as rinsing and blotting of the electrodes, response, stirring, and proper standardization. Besides these parameters associated with direct pH measurements, the potential measurements near the end point and the end-point determination have a large effect on accuracy. This includes the proper response and anticipation which are the result of a combination of factors including titrant strength and delivery rate, stirring rate, and electrode positioning. The objective is to obtain one-drop control near the end point in order that the largest potential change, the end point, can be easily detected. The addition of the drop should be sensed and a stable reading obtained before the next drop is added. Often the sharpness of the end point can be improved by adding smaller increments of higher strength titrant than normally used, since the titrant has less dilution effect on the measured potential.

6.2.8 Acid Fluoride Solutions

Glass electrodes exhibit erratic behavior in the presence of only $10^{-6}\, m$ hydrofluoric acid (HF) and therefore are not normally used below pH 6. Also, in low pH solutions containing HF, the glass electrode life is severely limited because of HF attack on the glass membrane.

There are two common methods for handling this difficult sample. The first is a direct method which limits the contact time between sample and electrodes. The second is an indirect method using a gold metallic electrode to measure the quinhydrone oxidation–reduction (redox) potential.

A. Direct Method

Direct pH measurement with a glass electrode can be made if only an approximate value is needed and reduced electrode life can be tolerated. If the measurement is taken quickly and the electrodes are thoroughly rinsed immediately after the measurement, the attack on the glass is limited. The erratic behavior previously mentioned is due to the formation of a new membrane hydration layer when the etching occurs. If the measurement is a screening value, such as to determine etching power for some plating baths, this behavior will not cause significant error.

The electrode life will be reduced by dissolving the glass and will first appear as reduced span when checking between two buffers. Continued use of this type of solution will eventually result in the inability to standardize the electrodes because of lack of sufficient standardization control on the meter. The growing span error may be minimized by using a standard buffer of pH value as close as possible to that of the sample.

There are two other versions of this method which are directed toward acidity in fluoride solutions. The first is to dilute the HF solution before measurement, and the second is a titration. If the only significant hydrogen complex in the solution is that of hydrogen fluoride and both reduced attack on the glass electrode and only approximate values were desired, a strongly acidic sample could be diluted with water. The hydrogen fluoride complex would not contribute hydrogen significantly below pH 3 ($pK_a = 3.17$ at 25°C), and the changes in pH due to ionic strength may be minor.

An acid–base titration can be performed on a fluoride solution by first adding a measured aliquot of standard base, which is insufficient to reach the end point, prior to immersing the glass electrode bulb in the sample. The titration will provide total acidity information which may be useful in some industrial application. This technique is best suited to a quality control application in which the end point is known and the quantity of base titrant varies over a narrow range.

B. Indirect Method

Indirect pH measurement using a gold metallic and a reference electrode measures the oxidation–reduction potential of quinhydrone which is pH dependent. If the sample is saturated with quinhydrone, an equimolar mixture of quinone and hydroquinone is established. Without discussing the details of the Nernst equation as done in Chapter 1, the relationship between the observed potential (E_0) using a calomel reference and the saturated sample is expressed by

$$E_0 = 455 - 59.2 \quad \text{pH} \qquad \text{at 25°C}$$

For example, a pH 4 buffer saturated with quinhydrone, which can be used to calibrate the system, would produce a potential of approximately 218 mV (455 − 237) at 25°C.

The major disadvantage of this technique is the interference of other oxidants or reductants. The pH glass bulb is very selective, whereas the metallic electrode is sensitive to all oxidation–reduction potentials in the sample. Solutions of high salt or acid content can also cause errors.

Although both methods are used to determine the pH of acid fluoride solutions, neither is entirely satisfactory, and selection becomes a matter of which method best fits the application.

6.2.9 Biological pH Measurements

There are a number of problems inherent in biological pH measurements. Biological or biomedical samples range from neutral and sterile to acidic gastric fluid. Often a great number of small-volume samples in small containers such as a test tube need to be measured. The sample may contain a high concentration of protein or be dry skin. Often high accuracy is required even though the sample pH may be changing with time, temperature, or exposure to air. The measurements are often made under anaerobic conditions at 37°C.

Most problems center about using the proper technique or equipment for the application. These factors have been discussed in detail in Chapters 2, 3, and 5. This section serves as a reminder of these factors.

The combination electrode is most often preferred in biomedical applications because of the ability to make measurements on limited sample size and/or sample container configurations. In some applications, however, the leakage of silver from the junction cannot be tolerated. If a pair of electrodes is used, the questions of proper reference junction flow rate and resistance should be considered.

The advantage in using a tris physiological buffer has already been discussed in Chapter 4. The most important advantage is compatibility with the sample so that the difference in liquid junction potential between buffer and sample is small.

Proper technique (discussed in Chapter 5) becomes very critical when attempting high accuracy measurement. For example, allowing the reading to stabilize, rinsing and blotting the electrodes without wiping, using a buffer of pH value close to that of the sample, and other steps are extremely important in obtaining a reliable answer for blood sample readings to a thousandth of a pH unit at pH 7.413.

For some applications, bacterial contamination by the electrodes must be prevented. Some manufacturers make electrodes which may be autoclavable. Other electrodes may be chemically sterilized with ethylene oxide or zephiran chloride. In either case, if the electrodes are standardized after the sterilizing procedure, the buffer must be sterile. If standardization occurs before autoclaving, a potential shift of about 10 mV or 0.2 pH unit can be expected.

Chapter 7

Troubleshooting

Problems which arise from malfunctioning equipment or from difficult sample measurements may be encountered. A general approach to problems encountered with difficult applications is discussed in Chapter 6, and determination by the preequilibrium test of which electrode is causing a slow response problem is presented in Chapter 5.

The objective of this chapter is to show how a malfunction of equipment can be isolated and corrected. Since there are only three parts to pH measuring equipment, the testing required to isolate the malfunction is limited. The sections in this chapter are arranged in the suggested order of testing, starting with the simplest and fastest test and progressing toward the more involved tests.

7.1   pH METER TEST

There is a simple test to verify that a pH meter is not grossly malfunctioning. It consists of shorting the inputs to divorce the meter from electrode behavior and then observing the reading as the controls are changed. This procedure does not test either the meter under high impedance conditions or the meter calibration as described in Chapter 2, but will result in detection of about 75% of meter malfunctions. Many manufacturers can supply a high-impedance shorting plug which provides additional assurance that the meter is being tested under conditions similar to operation with electrodes, or they can supply a terminal connector and shorting strap used to short the meter inputs.

The first step of this test, to short the glass and reference inputs, is done using a terminal connector and shorting strap. The terminal connector allows a pin jack to make connection with the glass input. The shorting strap with pin jacks at either end is connected between the terminal connector and the reference jack (see Figure 7.1). The meter is placed in the pH function, the temperature control at 30°C, and the slope control at 100% or off. If the meter has a zero control, it should be set to provide pH 7.00 on the display when the meter is in the stand-by mode. The readout is then activated to conduct the test.

The standardization control is rotated to its extreme positions while observing the display. If the span of the standardization control is ±100 mV, the readout will reflect greater than ±1.5 pH units change from pH 7.00; that is, the readout will display less than 5.5 pH and greater than 8.5 pH units. With a larger span on the standardization control, greater change will be observed. For example, a control with ±200 mV will exhibit greater than ±3 pH units from pH 7 (<4 to >10). The number of pH units change possible is indicated by the millivolt span of the standardization control divided by 60.

Then, set the display to read pH 6.00 with the standardization control. When the temperature compensating control is rotated to 0°C, the display should read about 5.9. When it is rotated to 100°C, it should read about 6.2. Reset the temperature control to 30°C and rotate the slope control. At 80% slope, the reading should be lower than pH 5.8.

By rotating or changing these controls, their relative effect on the reading can be observed, and any gross malfunction in these controls should also be observed. The time required to perform this

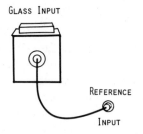

GLASS INPUT

REFERENCE

INPUT

Figure 7.1
Testing the pH Meter

test is about a minute, and the test provides a great deal of assurance that the malfunction is not caused by the meter. Figure 7.1 and the following steps outline this simple meter test.

(1) Short the inputs (see Figure 7.1).
(2) Set the controls at the starting point:
      temperature control, 30°C,
      slope control, 100%,
      zero control, pH 7.
(3) With the readout activated:

Rotate control	Initial pH	Observed pH readout
Standardization	7	±1.5
Temperature		
0°C	6	5.9
100°C	6	6.2
Slope		
80%	6	5.8

Some manufacturers supply a test function built into the meter. This function normally tests the analog circuit board and the readout circuit board. It provides a thorough test of the pH meter and can be used in place of the shorting test, and it eliminates the meter as a source of difficulty in a matter of seconds.

7.2   GLASS ELECTRODE TEST

The main criteria for a properly functioning glass electrode are response and span; that is, how fast the electrode responds, and how much pH change is observed when taking measurements of different pH value buffers. The main drawback of this test is the assumption that the reference electrode being used with the glass electrode is functioning properly. If this assumption is correct, a malfunctioning glass electrode can be detected by its short span and/or slow response to standard buffer solutions. For example, a glass bulb which has been scratched may exhibit short span, while a glass bulb which is coated with an oil film may exhibit slow response. These procedures are outlined in Section 3.1.9.

To review: the expected electrode response should reach 98% of its final reading within 10 seconds when the span test is performed between two buffers; the rejuvenation procedure, described in Figure 3.3, should be initiated if the glass electrode does not meet these specifications.

7.3   REFERENCE ELECTRODE TEST

As mentioned previously, the reference electrode is often a source of difficulty (see Figure 3.8). The difficulties may arise from a clogged or high resistance junction, a malfunctioning electrode, or a junction which is not compatible with the sample because of junction type or the filling solution.

A high liquid junction potential may arise from sample clogging, precipitation in the junction, or if the junction is allowed to dry. In order to determine if this is the source of difficulty, the junction *resistance test* described in Section 3.2.5.B may be implemented.

Another possible malfunction of the reference electrode is a broken internal. This is most obvious when a calomel internal is broken, thus causing the filling solution surrounding it to turn gray. The reference potential in this case is quite different from that of other calomel electrodes. The reference potential of an internal element which is just beginning to drift, however, is not as obvious. Either the obviously broken reference internal or the more subtle drifting potential can be confirmed by comparison with a properly functioning reference electrode, as outlined in the *bucking test* described in Section 3.2.5.A.

Since there is no universal junction which is compatible with or will perform well in all types of samples, one possible solution to difficulties may require the use of another reference electrode with a different type of junction. A general indication of preferred junctions for different types of samples was shown in Table 3.3. Most often, a sleeve junction will provide good performance and should be selected if the disadvantages of the flow rate and maintenance discussed in Chapter 3 are acceptable. Chapter 3 also discusses altering the reference filling solution to make the junction more compatible with the sample.

7.4   SYSTEM COMPARISON

Each previously described test has been intended to verify the operation of an individual component or relies on the assumption that another component is known to be functioning properly. There are some tests, however, which involve all the components of a pH measurement. One such test aids in identifying which electrode is causing a slow response. This preequilibration test is described in Section 5.3. Another test is a comparison with a second pH measuring system if it is available. This test involves cross-checking between measuring systems when they yield different pH values for the same solutions.

The purpose of this test is to pinpoint the source of deviation from two supposedly identical measuring systems. The comparison of total systems is only the first step, and if a difference is found, the source of deviation must be located. With two pairs or combination electrodes, a cross check by electrode substitution can be conducted. If electrode pairs (A) and (D) give different pH values for the same solution, there are two other possible electrode pairs (B) and (C), as shown in Figure 7.2.

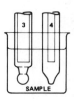

	Pairs	pH
(A)	1–2	7.75
(B)	1–4	7.68
(C)	3–2	7.75
(D)	3–4	7.68

(a)   Electrode pair cross-check example.

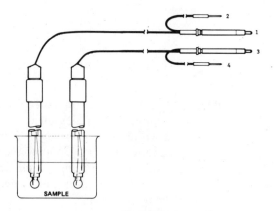

Connectors

	Glass	Reference	pH
(A)	1	2	7.75
(B)	1	4	7.68
(C)	3	2	7.75
(D)	3	4	7.68

(b)   Combination electrode cross check.

Figure 7.2
pH System Comparison

If only combination electrodes are available, they may be treated as pairs simply by connecting only that portion of the electrode under test. If, for example, pair (B) is being tested, the glass ferrule connector from the combination described as (1) and the reference pin jack connector from the combination described as (4) would be connected to the meter. The reference pin jack (2) would be connected to the solution ground terminal to provide a completed shielding circuit.

If the two pairs of electrodes show a pH difference in a sample of known pH value after being standardized on the same standard pH buffer solution, then the faulty electrode can be pinpointed. If the deviation is observed in a sample of unknown value, the problem can be traced only to either reference or either glass electrode. In other words, if pairs (A) and (C) and pairs (B) and (D) each agree on the pH value, the source of difficulty would be with one of the reference electrodes. If pairs (A) and (B) and pairs (C) and (D) each agree, the source of difficulty would be with one of the glass electrodes. At this point, individual electrode tests previously described in this chapter should be initiated to aid in specific identification of the faulty electrode.

In the case when the pH difference is noted in a buffer of known pH value, a faulty glass electrode would give a consistent offset from the known pH value in pairs (A) and (B) or (C) and (D). A faulty reference electrode would give a consistent offset in pairs (A) and (C) or (B) and (D).

These tests should be conducted on the same pH meter, but can be repeated on the second pH meter to eliminate the meter as a possible source of error and to add validity to the results. Of course, technique as described in Chapter 5 can seriously affect the results, and care should be taken to ensure that the measurements are made properly.

7.5 CONCLUSION

Troubleshooting becomes a matter of knowing how to locate the difficulty and then taking the proper action. The individual compo-

nent of a pH measuring system may be tested for performance, or
the entire system may be used to deduce the source of difficulty.
As stated previously, a troubleshooting procedure will often lead to
a reference electrode with a high liquid-junction potential. The next
steps to resolve this problem involve clearing the junction, changing the type of junction, or changing the filling solution composition. Some of these steps are discussed in Chapter 3.

If the troubleshooting procedure leads to a short-span or slow-
responding glass electrode, the rejuvenation of the glass bulb or
electrode replacement is required. If the problem is the result of a
faulty pH meter, only proper service will resolve the problem. No
matter which component is the cause of difficulty, the procedures
in this chapter should help to locate it.

Appendix

Tables of Data

TABLE A.1

Millivolts per pH Unit versus Temperature[a]

t (°C)	Slope $(RT \ln 10)/F$ (volts)
0	0.054197
5	0.055189
10	0.056181
15	0.057173
20	0.058165
25	0.059157
30	0.060149
35	0.061141
38	0.061737
40	0.062133
45	0.063126
50	0.064118
55	0.065110
60	0.066102
65	0.067094
70	0.068086
75	0.069078
80	0.070070
85	0.071062
90	0.072054
95	0.073046
100	0.074038

[a] R = 8.3143 joules degree^{-1} mole^{-1};
F = 96487 coulombs equivalent^{-1};
$T = t$ (°C) + 273.15;
$\ln 10 = 2.3026$.

TABLE A.2

Dielectric Constants[a]

Liquid	t	D
Acetic acid	20	6.15
Acetone	25	20.70
Acetonitrile	25	36.0
Ammonia	−77.7	25
Aniline	20	6.89
Benzene	25	2.274
Carbon tetrachloride	25	2.228
Chlorobenzene	25	5.621
Chloroform	20	4.806
Cyclohexane	25	2.015
Deuterium oxide	25	77.9
Dimethylformamide	25	36.7
Dimethylsulfoxide	25	46.6
1,4-Dioxane	25	2.209
Ethanol	25	24.3
Ethanolamine	25	37.7
Ethyl acetate	25	6.02
Ethyl ether	20	4.34
Ethylamine	10	6.94
Ethylenediamine	20	14.2
Formamide	20	109
Formic acid	16	58.5
Glycerol	25	42.5
Hydrazine	20	52.9
Hydrogen peroxide	0	84.2
Methanol	25	32.63
Methylamine	25	9.4
Methyl cellosolve	30	16.0
N-Methylacetamide	40	165
N-Methylpropionamide	25	176
Nitrobenzene	25	34.82
Nitromethane	30	35.87
Phosphorus trichloride	25	3.43
Pyridine	25	12.3
Sulfuric acid	25	101
Tin tetrachloride	20	2.87
Water	25	78.30

[a] Sources: Varied. See especially A. A. Maryott and E. R. Smith, "Table of Dielectric Constants of Pure Liquids." *NBS Circ.* 514 (August 10, 1951).

TABLE A.3

Clark and Lubs Buffers at 20°C

Composition	pH
48.5 ml 0.2 N HCl + 25 ml 0.2 N KCl diluted to 100 ml	1.0
32.25 ml	1.2
20.75 ml	1.4
13.15 ml	1.6
8.3 ml	1.8
5.3 ml	2.0
3.35 ml	2.2
46.70 ml 0.1 N HCl + 50 ml 0.1 M KHC$_8$H$_4$O$_4$ diluted to 100 ml	2.2
39.60 ml	2.4
32.95 ml	2.6
26.42 ml	2.8
20.32 ml	3.0
14.70 ml	3.2
9.90 ml	3.4
5.97 ml	3.6
2.63 ml	3.8
0.40 ml 0.1 N NaOH + 50 ml 0.1 M KHC$_8$H$_4$O$_4$ diluted to 100 ml	4.0
3.70 ml	4.2
7.50 ml	4.4
12.15 ml	4.6
17.70 ml	4.8
23.85 ml	5.0
29.95 ml	5.2
35.45 ml	5.4
39.85 ml	5.6
43.00 ml	5.8
45.45 ml	6.0

TABLE A.3

(continued)

Composition	pH
5.70 ml 0.1 N NaOH + 50 ml 0.1 M KH$_2$PO$_4$ diluted to 100 ml	6.0
8.60 ml	6.2
12.60 ml	6.4
17.80 ml	6.6
23.65 ml	6.8
29.63 ml	7.0
35.00 ml	7.2
39.50 ml	7.4
42.80 ml	7.6
45.20 ml	7.8
46.80 ml	8.0
2.61 ml 0.1 N NaOH + 50 ml 0.1 M H$_3$BO$_3$ diluted to 100 ml	7.8
3.97 ml	8.0
5.90 ml	8.2
8.50 ml	8.4
12.00 ml	8.6
16.30 ml	8.8
21.30 ml	9.0
26.70 ml	9.2
32.00 ml	9.4
36.85 ml	9.6
40.80 ml	9.8
43.90 ml	10.0

TABLE A.4

Solvent Conductivity

Liquid	Temperature (°C)	mhos/cm or ohm^{-1} cm^{-1}
Acetic acid	25	1.12×10^{-8}
Acetic anhydride	25	4.8×10^{-7}
Acetone	25	6×10^{-8}
Acetonitrile	20	7×10^{-6}
Acetophenone	25	6×10^{-9}
Acetyl chloride	25	4×10^{-7}
Allyl alcohol	25	7×10^{-6}
Ammonia	−79	1.3×10^{-7}
Aniline	25	2.4×10^{-8}
Benzaldehyde	25	1.5×10^{-7}
Benzene	—	7.6×10^{-8}
Benzonitrile	25	5×10^{-8}
Benzyl alcohol	25	1.8×10^{-6}
Isobutyl alcohol	25	8×10^{-8}
Capronitrile	25	3.7×10^{-6}
Carbon tetrachloride	18	4×10^{-18}
m-Chloraniline	25	5×10^{-8}
Chloroform	25	$<2 \times 10^{-8}$
Chlorohydrin	25	5×10^{-7}
Diethyl carbonate	25	1.7×10^{-8}
Diethyl oxalate	25	7.6×10^{-7}
Diethyl sulfate	25	2.6×10^{-7}
Ethyl alcohol	25	1.35×10^{-9}
Ethyl ether	25	$<4 \times 10^{-13}$
Ethyl nitrate	25	5.3×10^{-7}
Ethyl thiocyanate	25	1.2×10^{-6}
Ethylene chloride	25	3×10^{-8}

TABLE A.4

(continued)

Liquid	Temperature (°C)	mhos/cm or ohm^{-1} cm^{-1}
Formamide	25	4×10^{-6}
Formic acid	18	5.6×10^{-5}
Formic acid	25	6.4×10^{-5}
Furfural	25	1.5×10^{-6}
Glycerol	25	6.4×10^{-8}
Glycol	25	3×10^{-7}
Methyl acetate	25	3.4×10^{-6}
Methyl alcohol	18	4.4×10^{-7}
Methyl ethyl ketone	25	1×10^{-7}
Methyl nitrate	25	4.5×10^{-6}
Methyl thiocyanate	25	1.5×10^{-6}
Nitromethane	18	6×10^{-7}
Phenyl isothiocyanate	25	1.4×10^{-6}
Phosgene	25	7×10^{-9}
Phosphorus	25	4×10^{-7}
Phosphorus oxychloride	25	2.2×10^{-6}
Propionaldehyde	25	8.5×10^{-7}
n-Propyl alcohol	25	2×10^{-8}
Isopropyl alcohol	25	3.5×10^{-6}
Pyridine	18	5.3×10^{-8}
Quinoline	25	2.2×10^{-8}
Salicylaldehyde	25	1.6×10^{-7}
Sulfur dioxide	35	1.5×10^{-8}
Sulfuric acid	25	1×10^{-2}
Trichloroacetic acid	25	3×10^{-9}
Water	18	4×10^{-8}

TABLE A.5

Debye–Hückel Constants[a]

	(For ion-size parameters, *å*, in angstrom units)			
t	Unit volume of solvent		Unit weight of solvent	
(°C)	*A*	*B*	*A*	*B*
0	0.4918	0.3248	0.4918	0.3248
5	0.4952	0.3256	0.4952	0.3256
10	0.4989	0.3264	0.4988	0.3264
15	0.5028	0.3273	0.5026	0.3272
20	0.5070	0.3282	0.5066	0.3279
25	0.5115	0.3291	0.5108	0.3286
30	0.5161	0.3301	0.5150	0.3294
35	0.5211	0.3312	0.5196	0.3302
38	0.5242	0.3318	0.5224	0.3306
40	0.5262	0.3323	0.5242	0.3310
45	0.5317	0.3334	0.5291	0.3318
50	0.5373	0.3346	0.5341	0.3326
55	0.5432	0.3358	0.5393	0.3334
60	0.5494	0.3371	0.5448	0.3343
65	0.5558	0.3384	0.5504	0.3351
70	0.5625	0.3397	0.5562	0.3359
75	0.5695	0.3411	0.5623	0.3368
80	0.5767	0.3426	0.5685	0.3377
85	0.5842	0.3440	0.5750	0.3386
90	0.5920	0.3456	0.5817	0.3396
95	0.6001	0.3471	0.5886	0.3404
100	0.6086	0.3488	0.5958	0.3415

[a] From R. A. Robinson and R. H. Stokes, "Electrolyte Solutions," 2nd ed., p. 468, Academic Press, New York, 1959. The values for unit weight of solvent were obtained by multiplying the corresponding values for unit volume by the square root of the density of water at the appropriate temperature (see Glossary).

TABLE A.6

Buffer pH versus Temperature

t (°C)	Phthalate	Phosphate	Carbonate	Calcium hydroxide
0	4.003	6.984	10.317	13.423
5	3.999	6.951	10.245	13.207
10	3.998	6.923	10.179	13.003
15	3.999	6.900	10.118	12.810
20	4.002	6.881	10.062	12.627
25	4.008	6.865	10.012	12.454
30	4.015	6.853	9.966	12.289
35	4.024	6.844	9.925	12.133
40	4.035	6.838	9.889	12.043
45	4.047	6.834	9.856	11.841
50	4.060	6.833	9.828	11.705
55	4.075	6.834		11.574
60	4.091	6.836		11.449
70	4.126	6.845		
80	4.164	6.859		
90	4.205	6.877		
95	4.227	6.886		

TABLE A.7

Liquid Junction Potentials

Junction: Solution X | KCl (saturated)

Solution X	E_i
HCl, 1 M	14.1
HCl, 0.1 M	4.6
HCl, 0.01 M	3.0
HCl, 0.01 M; NaCl, 0.09 M	1.9
HCl, 0.01 M; KCl, 0.09 M	2.1
KCl, 0.1 M	1.8
$KH_3(C_2O_4)_2$, 0.1 M	3.8
$KH_3(C_2O_4)_2$, 0.05 M	3.3
$KH_3(C_2O_4)_2$, 0.01 M	3.0
KHC_2O_4, 0.1 M	2.5
KH phthalate, 0.05 M	2.6
KH_2 citrate, 0.1 M	2.7
KH_2 citrate, 0.02 M	2.9
CH_3COOH, 0.05 M; CH_3COONa, 0.05 M	2.4
CH_3COOH, 0.01 M; CH_3COONa, 0.01 M	3.1
KH_2PO_4, 0.025 M; Na_2HPO_4, 0.025 M	1.9
$NaHCO_3$, 0.025 M; Na_2CO_3, 0.025 M	1.8
Na_2CO_3, 0.025 M	2.0
Na_2CO_3, 0.01 M	2.4
Na_3PO_4, 0.01 M	1.8
NaOH, 0.01 M	2.3
NaOH, 0.05 M	0.7
NaOH, 0.1 M	0.4
NaOH, 1 M	8.6
KOH, 0.1 M	0.4
KOH, 1 M	6.9

Activity (a_i) A thermodynamic term for the apparent or active concentration of a free ion in solution. It is related to concentration by the activity coefficient.

Activity coefficient (f_i) A ratio of the activity of species $i(a_i)$ to its molality (C). It is a correction factor which makes the thermodynamic calculations correct. This factor is dependent on ionic strength, temperature, and other parameters.

  Individual ionic activity coefficients, f_+ for cation and f_- for an anion, cannot be derived thermodynamically. They can be calculated only by using the Debye–Hückel law for low concentration solutions in which the interionic forces depend primarily on charge, radius, and distribution of the ions and on the dielectric constant of the medium rather than on the chemical properties of the ions.

  Mean ionic activity coefficient ($f_\pm$) or the activity of a salt, on the other hand, can be measured by a variety of techniques such as freezing point depression and vapor pressure as well as paired sensing electrodes. It is the geometric mean of the individual ionic activity coefficients:

$$f_\pm = (f_+{}^{n+}f_-{}^{n-})^{1/n}$$

Anion A negatively charged ion (Cl^-, NO_3^-, S^{2-}, etc.)

Asymmetry potential   The potential developed across the glass membrane with identical solutions on both sides. Also a term used when comparing glass electrode potential in pH 7 buffer. (See Chapter 3.)

ATC   Automatic temperature compensation.

Bias current (input terminal current)   That part of the input current which is independent of the applied voltage from the electrode potentials.

Buffer   Any substance or combination of substances which, when dissolved in water, produces a solution which resists a change in its hydrogen ion concentration on the addition of an acid or alkali.

Buffer capacity (β)   A measure of the ability of the solution to resist pH change when a strong acid or base is added.

Calibration   The use of two or more standards, one to establish the standardization point and one to establish electrode sensitivity by a slope adjustment.

Cation   A positively charged ion (Na^+, H^+).

Conductance   The measure of the ability of a solution to carry an electrical current. (See Equivalent conductance.)

Debye–Hückel equation   Used in relating the activity coefficient (f_i) to ionic strength (see Activity coefficient):

$$-\log f_i = \frac{AZ_i^2 I^{1/2}}{1 + B\mathring{a}I^{1/2}}$$

where

I is the ionic strength,
A and B the temperature-dependent constants (see Table A.5),

Z_i the valence of the ion (i), and
å the ion-size parameter in angstroms.

Dielectric constant   Related to the force of attraction between two opposite charges separated by a distance in a uniform medium.

Dissociation constant (K)   A value which quantitatively expresses the extent to which a substance dissociates in solution. The smaller the value of K, the less dissociation of the species in solution. This value varies with temperature, ionic strength, and the nature of the solvent.

Electrode potential (E)   The difference in potential established between an electrode and a solution when the electrode is immersed in the solution.

Electrolyte   Any substance which, when in solution or fused, will conduct an electric current. Acids, bases, and salts are common electrolytes.

Electromotive force (emf)   The potential difference between the two electrodes in a cell. The cell emf is the cell voltage measured when no current is flowing through the cell. It can be measured by means of a pH meter with high input impedance.

End point (potentiometric)   The apparent equivalence point of a titration at which a relatively large potential change is observed.

Equilibrium constant   The product of the concentrations (or activities) of the substances produced at equilibrium in a chemical reaction divided by the product of concentrations of the reacting substances, each concentration raised to that power which is the coefficient of the substance in the chemical equation.

Equitransference   Equal diffusion rates of the positively and negatively charged ions of an electrolyte across a liquid junction without charge separation.

Equivalent conductance (λ)   Equivalent conductance of an electrolyte is defined as the conductance of a volume of solution containing one equivalent weight of dissolved substances when placed between two parallel electrodes 1 cm apart, and large enough to contain between them all of the solution. λ is never determined directly, but is calculated from the specific conductance (L_s). If C is the concentration of a solution in gram-equivalents per liter, then the concentration per cubic centimeter is $C/1000$, and the volume containing one equivalent of the solute is, therefore, $1000/C$. Since L_s is the conductance of a centimeter cube of the solution, the conductance of $1000/C$ cc, and hence λ, will be

$$\frac{1000L_s}{C}$$

Filling solution   A solution of defined composition to make contact between an internal element and a membrane or sample. The solution sealed inside a pH glass bulb is called an internal filling solution. This solution normally contains a buffered chloride solution to provide a stable potential and a designated zero potential point. The solution which surrounds the reference electrode internal and periodically requires replenishing is called the reference filling solution. It provides contact between the reference electrode internal and sample through a junction.

Hydrogen ion activity (a_{H^+})   Activity of the hydrogen ion in solution. Related to hydrogen ion concentration (C_{H^+}) by the activity coefficient for hydrogen (f_{H^+}).

Hysteresis (electrode memory)   When an electrode system is returned to a solution, equilibrium is usually not immediate. This phenomenon is often observed in electrodes that have been exposed to the other influences such as temperature, light, or polarization.

Input resistance (impedance)   The input resistance of a pH meter is the resistance between the glass electrode terminal and the reference electrode terminal. The potential of a pH-measuring electrode chain is always subject to a voltage division between the total electrode resistance and the input resistance.

Internal reference electrode (element)   The reference electrode placed internally in a glass electrode.

Ionic mobility   Defined similarly to the mobility of nonelectrolytic particles, viz., as the speed that the ion obtains in a given solvent when influenced by unit power.

Ionic strength   The weighted concentration of ions in solution, computed by multiplying the concentration of each ion in solution (C) by the corresponding square of the charge on the ion (Z), summing this product for all ions in solution and dividing by 2:

$$\text{ionic strength} = \tfrac{1}{2} \sum Z^2 C.$$

Isopotential point   A potential which is not affected by temperature changes. It is the pH value at which dE/dt for a given electrode pair is zero. Normally, for a glass electrode and SCE reference, this potential is obtained approximately when immersed in pH 7 buffer.

Liquid junction potential   The potential difference existing between a liquid–liquid boundary. The sign and size of this potential depends on the composition of the liquids and the type of junction used.

Mean ionic activity coefficient   See Activity coefficient.

Medium effect (f^m)   For solvents other than water the medium effect is the activity coefficient related to the standard state in water at zero concentration. It reflects differences in the electrostatic and chemical interactions of the ions with the molecules of various solvents. Solvation is the most significant interaction.

Membrane   The pH-sensitive glass bulb is the membrane across which the potential difference due to the formation of double layers with ion-exchange properties on the two swollen glass surfaces is developed. The membrane makes contact with and separates the internal element and filling solution from the sample solution.

Millivolt   Unit of electromotive force. It is the difference in potential required to make a current of 1 milliampere flow through a resistance of 1 ohm.

Molality   A measure of concentration expressed in mols per kilogram of solvent.

Molarity   A measure of concentration expressed in mols per liter of solution.

Monovalent ion   An ion with a single positive or negative charge (H^+, Cl^-).

Nernst equation   A mathematical description of electrode behavior:

$$E = E_x + \frac{2.3RT}{nF} \log a_i$$

E is the total potential, in millivolts, developed between the sensing and reference electrodes; E_x varies with the choice of electrodes, temperature, and pressure; $2.3RT/nF$ is the Nernst factor (R and F are constants, n is the charge on the ion, including sign, T is the temperature in degrees Kelvin), and a_i is the activity of the ion to which the electrode is responding.

Nernst factor (S, slope)   The term $2.3RT/nF$ is the Nernst equation, which is equal (at $T = 25°C$) to 59.16 mV when $n = 1$ and 29.58 mV when $n = 2$, and which includes the sign of the charge on the ion in the term n. The Nernst factor varies with temperature. (See Table A.1.)

Normal hydrogen electrode   A reversible hydrogen electrode (Pt) in contact with hydrogen gas at 1 atmosphere partial pressure and immersed in a solution containing hydrogen ions at unit activity.

Open circuit   The lack of electrical contact in any part of the measuring circuit. An open circuit is usually characterized by rapid large jumps in displayed potential, followed by an off-scale reading.

Operational pH   The determination of sample pH by relating to pH measurements in a primary standard solution. This relationship assumes that electrode errors such as sensitivity and changes in asymmetry potential can be disregarded or compensated for, provided the liquid junction potential remains constant between standard and sample.

pH(S) (standard pH scale)   The conventional standard pH scale established on the basis that an individual ionic activity coefficient can be calculated from the Debye–Hückel law for primary buffers.

Polarization   The inability of an electrode to reproduce a reading after a small electrical current has been passed through the membrane. Glass pH electrodes are especially prone to polarization errors caused by small currents flowing from the pH meter input circuit and from static electrical charges built up as the electrodes are removed from the sample solution, or when the electrodes are wiped.

Primary standards   Aqueous pH buffer solutions established by the National Bureau of Standards within the 2.5 to 11.5 pH range of ionic strength less than 0.1 and which provide stable liquid junction potential and uniformity of electrode sensitivity.

Redox potential   The potential developed by a metallic electrode when placed in a solution containing a species in two different oxidation states.

Salt bridge   The salt bridge of a reference electrode is that part of the electrode which contains the filling solution to establish the electrolytic connection between reference internal cell and the test solution.

Auxiliary salt bridge   A glass tube open at one end to receive intermediate electrolyte filling solution, and the reference electrode tip and a junction at the other end to make contact with the sample.

Salt effect (f^x)   The effect on the activity coefficient due to salts in the solution.

SCE   Saturated calomel electrode.

Secondary standard   pH buffer solutions which do not meet the requirements of primary standard solutions but provide coverage of the pH range not covered by primary standards. Used when the pH value of the primary standard is not close to the sample pH value.

Slope (electrode sensitivity, span)   See Nernst factor.

Solvation   Ions in solution are normally combined with at least one molecule of solvent. This phenomenon is termed solvation.

Standard electrode potential (E^0)   The standard potential $E^{0'}$ of an electrode is the reversible emf between the normal hydrogen electrode and the electrode with all components at unit activity.

Standardization   A process of equalizing electrode potentials in one standardizing solution (buffer) so that potentials developed in unknown solutions can be converted to pH values.

Suspension effect   The source of error due to varied reference liquid junction potential depending upon whether the electrodes are immersed in the supernatant fluid or deeper in the sediment. Normally encountered with solutions containing resins or charged colloids.

Zero point

pH meters The electrical zero point where zero millivolts would be displayed. Used in conjunction with the slope control to provide a narrower range calibration.

Electrode See Isopotential point.

Index

167

Q